ISBN-13: 9781463665012

ISBN-10: 1463665016

Cover: When a party of the last ape in our line of descent is marooned on a narrow, thinly treed littoral of Lake Mega Chad, females gather in a tight congregation alongside water so they can cooperate to protect their babies from male infanticide. In time, their expanding numbers oblige them to add mollusks from the Lake floor to their diet. To speed a female's dive, her pelvis shortens and rotates forward, creating hominin waist and buttocks.

From Toumai to
G. Stein and O. Wilde

Also by Arthur M. Squires

The Tender Ship: Governmental management of technological change, 1986.

Small Coal Furnaces: The neglected option, 1981.

Flue gas desulfurization: report of National Research Council prepared for Committee on Public Works, United States Senate, 1975.

With Mooson Kwauk and Amos Avidan

Fluid Beds: At Last, Challenging Two Entrenched Practices, *Science*, vol. 230, pp. 1329-1337, 1985.

With David A. Berkowitz (co-editors)

Power Generation and Environmental Change, 1971.

With Cuthbert Daniel

A Road to Atomic Peace, Christian Century, March-April 1949.

In preparation

The left hand of love: coevolution of humankind's curious sexualities and its equally strange ultra-sociality

Maestros of technology in WWII

Three historic fluid beds with lessons for today's practitioners

From Toumai to

G. Stein and O. Wilde

Arthur M. Squires

An unconventional story of hominin evolution from Toumai's species, *Sahelanthropus tchadensis*, at a shore of Lake Mega Chad to globe-straddling humankind and its curious sexualities: the typical woman's unawareness of an interval of fertility and a minority directing loves and lusts toward individuals of the same, not opposite, sex.

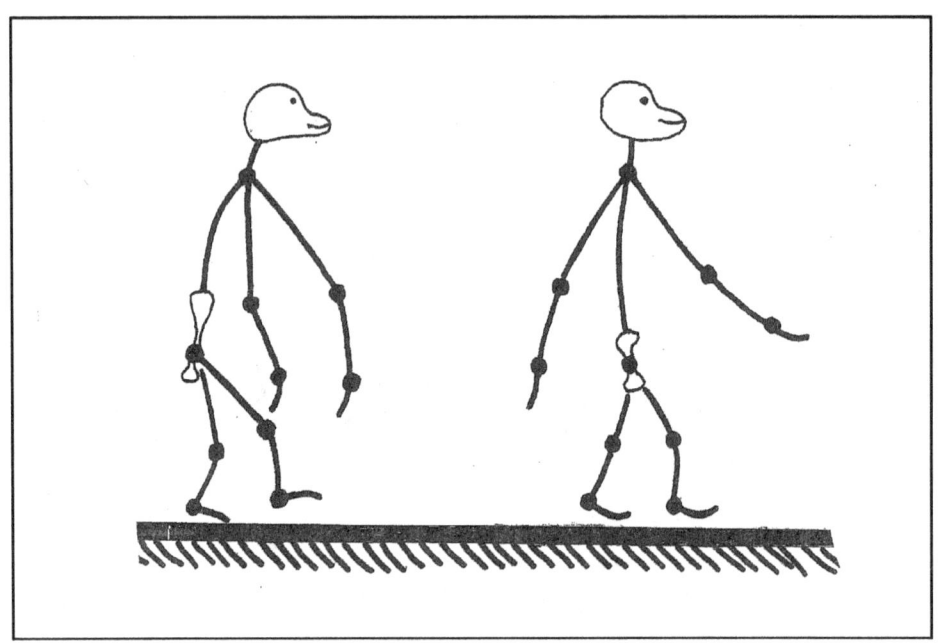

For the last male ape in our line, it's no big deal to walk on two feet; but for the male *S. tchadensis* it's easier, with his shortened, forwardly rotated pelvis and balanced head.

Contents

"[The human line's last ancestral ape species] *might have been very size dimorphic like most hominoids except chimpanzees, females weighing around 30 kg, males perhaps twice that.*"
David Pilbeam, 1980 (see reference 19).

Hominins in "Two Worlds"

How certain data in papers on* Ardipithecus ramidus *support, not startling hypotheses offered therein, but hypotheses re the first hominin female's counterstrategy to male infanticide and the male response thereto

Abstract

A profitably entertained hypothesis is that all known fossils of *Ardipithecus ramidus* are male in origin. Lack of *ramidus* female fossils supports a hypothesized lives-in-two-worlds sociosexual arrangement: males free to range widely, harvesting woodland foods; females, gathering in a tiny area alongside water, eating waterfoods, their fossils nearly impossible to find. This arrangement arises in *Sahelanthropus tchadensis* (Toumai's species) at a shore of Lake Mega Chad. A party of the last ape in our line of descent is stranded in an essentially one-dimensional environment: a narrow, fragmented stretch of woodland, water on one side, a proto-

1

desert on the other. Here, a female's chance of evading a murder-bent bachelor is much poorer than in the contiguous, three-dimensional woodland of her ancestors. She quickly learns to catch fish by hand, to dive for mollusks, to eat water-grass, and to cooperate in warding off a male stranger. Diving streamlines the hominin female's body, rotating and shortening the pelvis, creating hominin waist and buttocks, and moving foramen magnum forward, balancing the head. These changes oblige males to walk and trot on two feet. They cooperate in two buddy-groups: first, a band of insiders asserts monopoly of sexual access to a congregation of females; second, a band of outsiders bides its time, recruiting adolescent males whom insiders exile, gathering strength for a battle to scatter insiders and take over the congregation.

Startling hypotheses

Reporting its studies of an ample collection of *Ardipithecus ramidus* fossils,[1–11] a large wing of paleoanthropology offered startling hypotheses:

(a) that *ramidus*, a hominin living between ~4.9 and 4.2 million years ago (mya), exhibits small sexual dimorphism in

size of body — the wing having assumed its collection of
ramidus fossils to include examples of both sexes, yet seeing no
significant dichotomy in sizes of teeth or other structures;[1,5,6,11]

(b) that, similarly, for some millions of years, earlier
hominin species display small sexual dimorphism in size, and,
"likely" as well, so does the last chimpanzee/human common
ancestor (the CLCA);[1,11]

(c) that male/male conflict for sexual access to fertile
females, typical for mammalian species displaying large sexual
size dimorphism,[12,13] is absent in *ramidus* and in the earlier
species of (b), such absence being consistent with both size
monomorphism and significant reduction in canine size in
comparison with apes ancestral to the CLCA;[1,5,6,11]

(d) that males of *ramidus* and the earlier species of (b)
and (c) cooperate with females in child rearing and even
provide them with foods carried from a distance.[11]

Improbability of hypotheses
advanced in *Ardipithecus* papers

To my knowledge, no fossil evidence supports a hypothesis of
monomorphism in body size for the CLCA. To the contrary,

before initiations of orangutan, gorilla, and chimpanzee lines of descent, for some ten-odd millions of years closely related apes, thought or known to be significantly dimorphic in size,[14,15] populate Europe, Asia, and Africa: e.g., *Dryopithecus* and *Sivapithecus* (a.k.a. *Ramapithecus*, the much smaller female of *Sivapithecus*, initially regarded a separate, nearly size-monomorphic species). Orangutan and gorilla, two putative descendants of one of these apes, are highly dimorphic. As well, another putative descendant, *Lufengpithecus lufengensis*, living in China at some 7-odd mya, is so spectacularly dimorphic[16] that its fossil remains, like *Sivapithecus* and *Ramapithecus*, were initially thought to represent two species, one small and one large.

Furthermore, australopithecines, putative descendants of *ramidus*, are sexually dimorphic in body size.[17] In the view set forth in the *ramidus* papers, this dimorphism is a reversion to a condition obtaining in our line millions of years earlier; the papers barely mention this improbable reversion[1,11] and offer no reason for it.

In addition, shrinkage of early hominin canines does not necessarily imply reduction in male/male conflict. This change may simply accompany bipedality, a concomitant of the foramen magnum's forward placement and the jaw's widening.[18]

After speciation of *tchadensis*, some five-odd million years must pass before hominin sociosexual intercourse evolves to a pattern resembling that which one of the *ramidus* papers[11] assigns as "likely" obtaining in the CLCA. The startling nature of this assignment, as well as of other hypotheses advanced in the *ramidus* papers, places paleoscientists under an obligation to seek and at least entertain alternative explanations of the *ramidus* data.

All *Ardipithecus* fossils are male

A candidate explanation is that **all** known *ramidus* fossils are remains of males. The body size is about right: the individual represented by a reconstructed skeleton (specimen ARA-VP-6/500) is thought to have weighed about 50 kilograms when alive,[1,10] not far below David Pilbeam's estimate[19] of 60 kg for the male CLCA; his estimate for the CLCA female is 30 kg. Male australopithecines weigh between about 40 and 49 kg; females are smaller by about 21% to 34%.[17] Lucy's "big brother," recently discovered, is ~30% larger than Lucy.[20]

Paleoanthropology does not have a stellar record for the accurate identification of the sex of the individual represented by a set of hominin bones. The discipline has been obliged to alter long believed, published identifications. For at least two decades, leading paleoanthropologists thought *Ramapithecus* to be a hominin ancestor of ours.[21] They thought it a monomorphic species. They thought their collection of its fossils included bones of both male and female origin. At last, in the early 1980s, discovery of additional fossils forced a retraction. It became clear that *Ramapithecus* was simply the female of *Sivapithecus* (which paleoscience had also thought to be a monomorphic spcies, the collection of its fossils representing individuals female as well as male).

The resemblance of the monomorphic character of the known fossils of *A. ramidus* to the long held, mistaken view of *Ramapithecus* and *Sivapithecus* should trouble paleoscience. Has this discipline found **any** set of fossils it can with complete certainty identify as representing a hominin species monomorphic in body size? I am not aware of such.

That all known *ramidus* fossils are male is consistent with (supportive of) the hypothesis[22] that early hominin females and males live in separate "worlds": a number of females gather in a tight congregation occupying a short, narrow strip of land alongside a river or a lake; their diet comprises, primarily, aquatic foodstuffs available in reliable abundance, season to season, year to year.[23] A smaller number of mature males, cooperating in a buddy-group guard, range over a significantly larger area of land nearby the female congregation; their diet comprises woodland foods familiar to their male CLCA ancestors. This guard of "insiders" zealously maintains its monopoly of sexual access to "their" females, brooking no competition from a junior male just achieving sexual maturity, whom they consign to an outback where, with luck, he manages to join a band of "outsiders" biding its time until it has gathered sufficient strength to challenge the insider guard to a battle, killing or scattering its members and assuming control of its females. Since both insider and outsider males travel freely, their fossil remains are widely dispersed; in contrast, female fossils are concentrated in tiny areas, to be found only by an exceedingly lucky paleoscientist.

In brief, the separate-world hypothesis assumes male exogamy and non-kin male coalitions warring for sexual access to stay-at-home congregations of female kin. A junior female is little inclined to wander; in possession of valuable knowledge of the aquatic foods of her home scene, she is loath to chance an unfamiliar outside. At 7-odd mya, the separate-world social arrangement arises in the first member of the hominin line.

Toumai and Lake Mega Chad

Michel Brunet flouted paleoanthropology's wisdom by making a surprising (and brave) commitment to a search for early hominin fossils in Chad's central Djurab region, some 1500-odd miles west of the Rift Valley, the orthodox location for such a search. The logic of central Chad as a place for Brunet's search comes from presence of a large, ancient, fresh-water lake, Lake Mega Chad, whose shores expose sedimentary rock of the same age as (and older than) strata of the Rift Valley.[24] In 1995, Brunet startled paleoscience by announcing discovery of an australopithecine (an approximate contemporary of Lucy);[25] and in 2002 came an even greater surprise, Brunet's discovery of the skull of "Toumai," who lives sometime between ~7.4 and 6.5 mya.[26,27] Brunet dubbed Toumai's species *Sahelanthropus tchadensis*. Some paleoscientists rejected *tchadensis* as hominin,[28] but a *ramidus* paper[5] confirms its hominin status, the oldest known to science.

Surrounding Lake Mega Chad is sandy desert, a proto-Sahara; the lake's source of water is storm runoff; no great river enters the lake, making for easy travel to the mid-African jungle and woodland.[24] As the lake forms, nearby bands of the CLCA are "marooned," their contact with the bulk of Africa's ape population broken. The lake's shoreline, extending some 2,300-odd miles, presents the CLCA welcoming habitats of many kinds, as well as repellent stretches of deserts, cliffs, arroyos, and swamps. Bands of CLCA tend to occupy "oases" — stretches of forest not at first much different from the great African forest beyond the proto-Sahara. At ~6.5 mya, however, African grasslands are expanding, forests are fragmenting.[29] An ape refugee prefers a forested oasis to a spot that lacks its familiar woodland foods. Where an oasis is wide and relatively continuous, a band of apes experiences little (if any) evolutionary pressure for change; its home scene is not an immediate venue for speciation (but, we will see, in time becomes a venue for initiation of the chimpanzee line). On the other hand, some forested stretches of shoreline are narrow and fragmented. These are venues for speciation of *tchadensis*, an outcome of a CLCA female counterstrategy against male infanticide.

Male infanticide and female counterstrategies

Substantially throughout the mammalian Order, a hazard for females is loss of a nursing infant to a murder-bent bachelor.[30] A common response is female cooperation to resist the danger, e.g., the pride of kin female lions protecting kits from infanticidal, non-kin males.[31] Similar behavior has been seen in barnyard cat and monkey mothers. In the sociosexual arrangement, harem-master and harem, the harem-master's job (e.g., the silver-back gorilla) is to protect nursing infants in arms of females in his charge.

If he fails, a dead infant's mother, often, within days, becomes fertile and presents herself to the murderer for impregnation. Sarah Hrdy offered an explanation.[32–35] In the male, evolution has selected for an action that earns him the earliest possible opportunity to sire an offspring. In the female, evolution has selected for the fertility that follows her loss, permitting her to cooperate in the male's design. Her response is realistic. He is now, and will probably remain, the closest competent, mature male at hand, the best available protector of her next infant.

A male gorilla, upon overturning a silverback, kills all infants under about two; and he kills a new baby born within a few months. Yet evolution has provided a female early in pregnancy a trick that will protect her unborn infant: she displays false signs of fertility and present herself for sex. Her baby will be safe, since evolution has not equipped the new silverback's brain with means for judging time to birth, if the time is sufficient.

Female counterstrategy to male infanticide at Lake Mega Chad

The hominin separate-world social arrangement arises as a natural consequence of the maroonment of a small number of CLCA in a narrow strip of fragmenting woodland alongside Lake Mega Chad. The environment is effectively one-dimensional, placing a nursing female at great hazard from a bachelor bent upon killing her infant. In her ancestor's three-dimensional, contiguous woodland, a female has a far better chance for escape: she can race away, either on the ground or in the tree canopy, in a direction the bachelor can only guess at. At Lake Mega Chad, she must find new foods, substitutes for the land's produce, whose harvest is now too dangerous. Lake Mega Chad is home to aquatic life adapted for living in both well-oxygenated deep waters and poorly aerated, mucky soups.[24] CLCA females will find foods in either kind of water; they will seek out those special places that best meet their needs (taking no heed for mature male well-being).

Coming together at the waterside furnishes CLCA females a counterstrategy to male infanticide. Cooperating, swimming females can readily prevent a bachelor from penetrating their circle, where they sequester their infants. Just three Rhine Maidens, in their element, can deal with thievish Hagen. CLCA females capture fish by hand,[36] and parties of females swim along their shore to exploit beds of shellfish and thickets of water-grass.

As Sarah Hrdy pointed out,[37] male breeding strategies respond to "how females space themselves." Whatever the CLCA male's historic breeding strategy — he probably is a harem-master monopolizing sexual access to a relatively small harem — he now shares a need with other nearby males: survival of each male's offspring. This need fosters evolutionary selection of males who cooperate with one another in a buddy-group alliance arraying itself in a guard that protects from infanticide all infants born to "their" females — and, as well, maintains a monopoly of sexual access to these females.

Promiscuity is a feature of female counterstrategy against male infanticide; it allows each male in her guard to suppose he has fathered her nursing infant. On her initiative, she can afford each male frequent opportunity for sexual congress. Unlike an ovulating chimpanzee female, she does not advertise her condition. Even if the CLCA (like the male gorilla[38]) can detect his consort's fertility through odor, this historic sign is hidden so long as the *tchadensis* female remains in the water. Males quickly respond to other signs of invitation. A signaling female invites a male to follow her along the shore to a spot where another male, whose jealousy their coupling might arouse, cannot see it. A curious feature of human sexual behavior is a desire for privacy; most mammalian sex occurs in public. Secrecy increases the *tchadensis* female's opportunity, when fertile, to select the father of her next child. A female treading water signals her intentions to a male on land. Two human strangers, distant from one another in a crowded room, instantly appreciate an eye signal inviting sex: eye contact, maintained for a mere instant too long, can indicate both are willing. Our quick understanding of an eye signal is a relic of the diving female's repertoire of signs inviting a male for sex.

The white sclera surrounding the human eye's colored iris may well appear early at Lake Mega Chad. If so, the whites of a *tchadensis* female's eyes reveal, to just one male target, instantly and unmistakably, the direction of her gaze. Today, these whites survive in us as a rare (unique?) feature in mammalian morphology.[39,40] Evolution of our line's ultra-sociality finds another use for our sclera: we are more generous when under the gaze of a human eye — even an artificial eye on a placard or at the top of a totem pole.

Diving females and trotting males

Although at first a CLCA female congregation at Lake Mega Chad can gather sufficient foods in shallow waters, expanding population obliges its members to dive for mollusks in deeper waters. In all sea mammals where a pelvis survives, the pelvis has rotated, streamlining their bodies. In our diving female ancestor, the pelvis rotates and shortens, creating hominin waist and buttocks. Streamlining cuts precious seconds from her travel to lake-bottom and back to surface for breath. Her streamlining creates *Sahelanthropus tchadensis*.

For the diving female's brother, upright walking evolves from necessity. Although a standing posture may at first give the CLCA male a bit of trouble, he is soon able to cover ground in strides as long as his short legs allow. He finds trotting a better way to cover ground at at least moderate speed than whatever he can manage on all fours. For both male stride and female streamlining, it helps when the hominin head moves into a straight line above pelvis and feet — i.e., when the foramen magnum moves forward to a central position at skull bottom.

Yet upright[41] has its downside: evolution concocts for the modern human a system full of aches, pains, and difficulties. As Gordon Hewes remarked,[42] "The persistence of inguinal hernia, intervertebral disc difficulties, various circulatory defects, etc. . . . suggests that natural selection for bipedalism among the emergent [hominins] must have been rapid and ruthless." Congregating and diving is, for *tchadensis* females, an urgent necessity.

For students of the human story, the why of our erect stature remains a puzzle. Elaine Morgan[43] wrote: "dozens of hypotheses about bipedalism have been found convincing enough to be accepted and published by the peer-review journals over the years"; with good humor, she disposed of six hypotheses still current. Citing behaviors of bonobo and proboscis monkey, she offered her own hypothesis: wading in shallow waters encourages, at times even requires, bipedal locomotion. Yet, although these species (as well as the lowland gorilla) have long occupied swamps, their pelvises have not rotated forward; when upright, their heads are not balanced.

As in most mammals, CLCA female and male look much alike (although the female is much the smaller). In contrast, women and men are strikingly different, in body shape, in quantity and disposition of fat, in physical strength. In the mammalian Order such large differences are rare. In the human female, a curious feature is how much of her fat appears in her buttocks. Could this serve as a keel to help her float face upward late in a pregnancy? An interesting experiment would be to recruit women who, wishing to remain thin, have starved themselves during a pregnancy, and find out how well they can float or swim when parturition is near.

War at waterside

Possessing special knowledge of the local lake bottom, a *tchadensis* female adolescent is disinclined to leave her mother's lakeside home. Her buddy-group guard welcomes her continued presence, since she presents a new opportunity for fathering. In contrast, what a guard sees in an adolescent male is an unwelcome rival. A cadet must take his chances on the outside. Since he can expect no welcome from a nearby guard, he is obliged to seek and hope to enter a band of bachelor outsiders.

In *tchadensis*, warfare between outsider and insider male bands is endemic.

(In a magisterial review of literature on origins of warfare,[44] Johan van der Dennen also suggested that an aim of chronic warfare among coalitions of early hominin males is to win reproductive access to females. His suggestion connects, of course, with Freud's "scientific myth" of the "primal horde" and its goals, castrating the father and capturing his women.[45,46])

Tchadensis outsiders gather recruits until they achieve strengths sufficient, in number and maturity, to challenge an insider group. A well-led outsider band bides its time until it is strong enough to attack, defeating insiders, killing some, exiling others to the outback. There, the latter, if their numbers are sufficient, found and lead a new band; or they may be welcomed by a band of young outsiders who appreciate the political advice the oldsters can offer.

For the hominin line, warfare may serve a biological purpose. At intervals, it sends fathers into exile and brings younger, virgin males onto the reproductive scene. In Japanese macaques, a male tends to depart a troop after ~5 years stay, emigrating to join another troop. As Sarah Hrdy surmised, this behavior's biological importance may be to lower incidence of incest between daughter and father.[47] For early hominins, warfare may have a similar underlying significance.

In many species, male canine teeth are primarily instruments for displays that assert dominance. The hominin male's unusually short and blunt canine teeth suggest that a dramatic display becomes less important for an alpha male than political savvy and successful generalship during battle. As I have mentioned, reduction in early hominin canine teeth size is likely a concomitant of bipedality,[18] yet if early hominin warfare were merely to comprise a number of one-on-one engagements, sharp canines, becoming useful, might quickly reappear through natural selection. Early hominin warfare, I suggest, involves more than single combat. An attacking party arms itself with sharp rocks; modifying a rock to provide a sharp edge is, perhaps, a first step toward invention of tools for cutting flesh. Since a worthy insider alpha male will post guards, a battle likely involves two parties hurling rocks at each other. Does any species other than humankind engage in massed warfare at a distance?

Tchadensis probably displays the three attributes[48,49] enabling the modern human to hurl an object with speed and precision: (1) likely inherited from the CLCA, ability to bend the hand back at close to a right angle relative to forearm;[1] (2) like *ramidus*, ability to grasp the object firmly;[7] and (3) ability to snap the hand forward at the moment of release of the thrown object. I suggest hominins acquire these attributes long before they invent their first stone tool, for whose production the three abilities are also important.[48] I cannot doubt that a female *tchadensis* is as bright as the sea otter, soon learning how to open two mollusks by striking them together while floating on her back. Back-bending, firmly grasping hand and ability to snap it forward smartly meet her need to process and consume foods she brings up from water-bottom without being obliged to return to shore. In *tchadensis* males, throwing improves in both distance and aim as they acquire hand skills that evolve initially to meet female necessity.

Initiation of the chimpanzee line

Outsider *tchadensis* bands, while gathering strength sufficient to challenge an insider guard, maraud at Lake Mega Chad's shores, where other marooned CLCA occupy wide, effectively three-dimensional littorals that closely resemble habitats familiar to the CLCA and its immediate ancestors. At these littorals, *tchadensis* males rape CLCA females and kill their babies, thereby initiating evolution of the chimpanzee line: male CLCA kin respond by cooperating in a guard that corrals non-kin females in an effort to protect their infants and prevent *tchadensis*/proto-chimpanzee couplings. For the latter aim, the proto-chimpanzee guard is not always successful: for several million years, such couplings produce viable offspring.[50]

Emigrants from Lake Mega Chad

Population pressure obliges parties of hominins to leave Lake Mega Chad. Female preference for living in a congregation alongside water is decisive in a party's selection of a new home. The ancient geography of Africa's Rift Valley is such that wet habitats are nearby sites of hominin fossil discoveries (including the stretch of woodland, seen in a six-mile-long outcrop at Aramis, Ethiopia, where was found the large sample of *Ardipithecus*[1]). Often, as well, the discovery scene itself is "wet": adjacent an ancient water, lake, river, or seashore.[51–54] Females and males continue to occupy separate worlds[22] until *Homo* appears.

References

1.   White T.D., Asfaw B., Beyene Y., *et al.* "*Ardipithecus ramidus* and the Paleobiology of Early Hominids." *Science.* 326:64, 75-86. 2009.

2.   WoldeGabriel G., Ambrose S.H., Barboni D., *et al.* "The Geological, Isotopic, Botanical, Invertebrate, and Lower Vertebrate Surroundings of *Ardipithecus ramidus.*"*Science.* 326:65. 2009.

3.   Louchart A., Wesselman H., Blumenschine R.J., *et al.* "Taphonomic, Avian, and Small-Vertebrate Indicators of *Ardipithecus ramidus* Habitat." *Science.* 326:66. 2009.

4.   White T.D., Ambrose S.H., Suwa G., *et al.* "Macrovertebrate Paleontology and the Pliocene Habitat of *Ardipithecus ramidus.*" *Science.* 326:87-93. 2009.

5.   Suwa G., Asfaw B., Kono R.T., *et al.* "The *Ardipithecus ramidus* Skull and Its Implications for Hominid Origins." *Science.* 326:68. 2009.

6.   Suwa G., Kono R.T., Simpson S.W., Asfaw B., *et al.* "Paleobiological Implications of the *Ardipithecus ramidus* Dentition." *Science.* 326:69, 94-99. 2009

7.   Lovejoy C.O., Simpson S.W., White T.D., *et al.* "Careful Climbing in the Miocene: The Forelimbs of *Ardipithecus ramidus* and Humans Are Primitive." *Science.* 326:70. 2009.

8.   Lovejoy C.O., Suwa G., Spurlock, L., *et al.* "The Pelvis and Femur of *Ardipithecus ramidus*: The Emergence of Upright Walking." *Science.* 326:71. 2009.

9.   Lovejoy C.O., Latimer B., Suwa G., *et al.* "Combining Prehension and Propulsion: The Foot of *Ardipithecus ramidus.*" *Science.* 326:72. 2009.

10.   Lovejoy C.O., Suwa G., Simpson S.W., *et al.* "The Great Divides: *Ardipithecus ramidus* Reveals the Postcrania of Our Last Common Ancestors with African Apes." *Science.* 326:73, 100-106. 2009.

11.   Lovejoy C.O. "Reeaximining Human Origins in Light of *Ardipithecus ramidus.*" *Science.* 326:74. 2009.

12.   Clutton-Brock T.H., Harvey P.H. "Primate Ecology and Social Organisation." *J Zool London.* 183:1-39. 1977.

13.   Alexander R.D., Hoogland J.L., Howard R.D., *et al.* "Sexual dimorphisms and breeding systems in pinnipeds, ungulates, primates, and humans." In Chagnon N.A., Irons W. *Evolutionary Biology and Human Social Behavior: An Anthropological Perspective.* North Scituate MA: Duxbury Press. 1979. pp. 402-435.

14.   Pickford M.H.L. "Geology, palaeoenvironments and vertbrate faunas of the mid-Miocene Ngorora Formation, Kenya." In Bishop

W.W. *Geological Background in Fossil Man.* Edinburgh, Scotland: Scottish Academic Press. 1978. pp. 237-262.

15. Harrison T. "Apes Among the Tangled Branches of Human Origins." *Science.* 327:532-534. 2010.

16. Kelley J., Xu Q. "Extreme sexual dimorphism in a Miocene hominoid." *Nature.* 352:151-152. 1991.

17. Aiello L.C. "Variable but singular." *Nature.* 368:399-400. 1994.

18. Mills J.R.E. "Occlusion and malocclusion in primates." In Brothwell D. *Dental Anthropology.* Oxford: Pergamon, 1963. pp. 29-51.

19. Pilbeam D. "Major Trends in Human Evolution." In Konigsson L.K. *Current Argument on Early Man.* Oxford: Pergamon. 1980. pp. 261-285.

20. Haile-Selassie V., Latimer B.M., Alene M., *et al.* "An early *Australopithecus afaremsis* postcranium from Roanso-Mille, Ethiopia." *Proc Natl Acad Sci USA.* 107:12121-12126. 2010.

21. Simons E.L. "Ramapithecus." *Sci Am.* 236(5):28-35. 1977.

22. Squires A.M. "Origins of Male In-group Behaviors." In Thienpont K., Cliquet R. *In-group/Out-group Behaviour in Modern Societies: An Evolutionary Perspective.* Brussels, Belgium: NIDI CBGS Publications. 1999. pp. 109-135.

23. Yesner D.R. "Maritime Hunter-Gatherers: Ecology and Prehistory." *Curr Anthropol.* 21:727-750. 1980.

24. Vignaud P., Duringer P., Mackaye H.T., *et al.* "Geology

and palaeontology of the Upper Miocene Toros-Menalia hominid locality, Djurab Desert, Northern Chad." *Nature.* 418:152-155. 2002.

25.   Brunet M., Beauvilain A., Coppens Y., *et al.* "The first Australopithecine 2,500 kilometres west of the Rift Valley (Chad)." *Nature.* 378:273-275. 1995.

26.   Brunet M., Guy F., Pilbeam D., *et al.* "A new hominid from the Upper Miocene of Chad, Central Africa." *Nature.* 418:145-151. 2002.

27.   Brunet M., Guy F., Pilbeam D., *et al.* "New material of the earliest hominid from the Upper Miocene of Chad." *Nature.* 434:752-755. 2005.

28.   Wolpoff M.H., Hawks J., Senut B., *et al.* "An Ape or *the* Ape: Is the Toumaï Cranium TM 266 a Hominid?" *PaleoAnthropology.* 2006:36-50. 2006.

29.   Brain C.K. "The evolution of man in Africa: Was it a consequence of Cainozoic cooling?" Alex L. du Toit Memorial Lecture No. 17. Annexure, Geological Society of South Africa. 85:1-19. 1981.

30.   Hausfater G., Hrdy S.B. *Infanticide: Comparative and Evolutionary Perspectives.* New York: Aldine. 1984.

31.   Pack C., Pusey A.E. "Divided We Fall: Cooperation among Lions." *Sci Am.* 276(5):52-59. 1997.

32. Hrdy S.B. "Care and Exploitation of Nonhuman Primate Infants by Conspecifics other than the Mother." *Advances in the Study of Behavior* 6:101-158. 1976.

33. Hrdy S.B. "Infanticide among Animals: A Review, Classification, and Examination of the Implications for the Reproductive Strategies of Females." *Ethol Sociobiol.* 1:13-40. 1979.

34. Struhsaker T. T., Leland L. "Colobines: Infanticide by Adult Males." In Smuts B.B., Cheney D.L., Seyfarth R.M., Wrangham R.W., Struhsaker T.T. *Primate Societies*. Chicago IL: Univ. of Chicago Press. 1987. pp. 83-97.

35. Sommer V. "Infanticide among Langurs of Jodhpur: Testing the Sexual Selection Hypothesis with a Longterm Record." In Parmigiani S., von Saal F.S. *Infanticide and Parental Care.* Newark NJ: Harwood Academic Publishers, Gordon & Breach. 1994. pp. 155-198.

36. Stewart K.M. "Early Hominid Utilisation of Fish Resources and Implications for Seasonality and Behaviour." *J Hum Evol.* 27:229-245. 1994.

37. Hrdy S.B. *The Woman That Never Evolved*. Cambridge MA: Harvard Univ. Press. 1981. p. 95.

38. Manson W.C. "Sexual Cyclicity and Concealed Ovulation." *J Hum Evol.* 15:21-30. 1986.

39. Emery N.J. "The eyes have it: The neuroethology, function and evolution of social gaze." *Neurosci Biobehav Rev.* 24:581-604. 2000.

40.   Milinski M., Rockenbach B. "Spying on Others Evolves."
        Science. 17:464-465. 2007.

41.   Ackerman J. "The downside of Upright: All those aching backs
        may be trying to tell us something: It's part of the price we pay for
        walking on two legs." *Nat Geog.* 210(1):126-145. 2006.

42.   Hewes G.W. "Food Transport and the Origin of Hominid
        Bipedalism." *Am Anthropol.* 63:687-710. 1961.

43.   Morgan E. *The Aquatic Ape Hypothesis.* London: Souvenir Press.
        1997. p. 52.

44.   van der Dennen J.M.G. *The Origin of War: The Evolution of a
        Male-coalitional Reproductive Strategy.* Groningen, The
        Netherlands: Origin Press. 1995.

45.   Freud S. *Totem and Taboo.* New York: Brill. 1916.

46.   Fox R. *The Red Lamp of Incest.* New York: Dutton. 1988.

47.   Hrdy S.B. "Raising Darwin's Consciousness: Female Sexuality
        and the Prehominid Origin of Patriarchy." *Human Nature.* 8:1-49.
        1997.

48.   Schick K.D., Toth N. *Making Silent Stones Speak.* New York:
        Simon & Schuster. 1993. pp. 135-139.

49.   Savage-Rumbaugh E.S. "Hominid Evolution: Looking to Modern
        Apes for Clues." In Quaitt D., Junichiro I. *Hominid Culture in
        Primate Perspective.* Niwot CO: Univ. of Colorado Press. 1994.
        pp. 7-49.

50.   Patterson N., Richter D.J., Gnerre S., *et al.* "Genetic evidence for complex speciation of humans and chimpanzees." *Nature.* 441:1103-1108. 2006.

51.   Wood B., Strait D. "Patterns of resource use in early *Homo* and *Paranthropus.*" *J Hum Evol.* 46:119-162. 2004.

52.   Puech P.-F., Albertini H., Serratricel C. "Tooth Microwear and Dietary Patterns in Early Hominids from Laetoli, Hadar and Olduvai." *J Hum Evol.* 12:721-729. 1983.

53.   Tobias P.V. 1998. "Water & Human Evolution." *Out There.* 38:35, 38-44. 1998. Available on line at http//archive.outthere.co.za/98/dec98/disp1.dec.html.

54.   Blumenschine R.J., Peters C.R., Masao F.T., *et al.* "Late Pliocene *Homo* and Hominid Land Use from Western Olduvai Gorge, Tanzania." *Science.* 299:1217-1221. 2003.

"Perhaps the greatest contrast between terrestrial and marine foods lies in the much greater spatial and temporal reliability of the latter. . . . There is …very little evidence of nutritionally related disease among coastal groups." David R. Yesner (see reference 23).

Australopithecus to *Homo*

When the woman becomes ignorant of her interval of fertility, life span increases and minding of babies by older offspring and her mother allows her to space her pregnancies at shorter intervals

Abstract

An australopithecine female is freer to travel about than the female *Ardipithecus ramidus*, but the lives-in-two-worlds sociosexual arrangement of *Sahelanthropus* persists into the early *Homo* line, a diving mother trading the minding of her baby with a protolesbian lover. The senior male's impulse to selectively foster cadets of gentle, "domesticatable" disposition, together with a female preference for cadets so fostered, accounts for neotenic evolution's major role in shaping hominin morphologies and behaviors. A crisis in the *Homo* story occurs when women become ignorant of their interval of fertility; protolesbians become mothers, putting pay to their ability or desire to help lovers. In response, years of childhood expand as well as years beyond menopause, and baby minding by older offspring and grandmothers provides new helps for mothers.

A female-female pair bond

When a female of the last chimpanzee/human common ancestor (the CLCA), dwelling at a narrow, fragmented littoral of Lake Mega Chad, becomes obliged to dive for food, picture her dilemma: how can she ensure her offspring's safety while she browses on shellfish from the deep? In her, no behavioral patterns are in place to produce female cooperation. Although CLCA females quickly learn to fight alongside one another against an intruding male stranger, in their mothering, each must at first fend for herself.

Although easily harvested shore foods meet the needs of the first few CLCA females marooned alongside Lake Mega Chad, in time an increase in their congregation's numbers produces an effect equivalent to food scarcity. A rank order in the congregation having evolved, a mother of high rank recruits help by an unattractive behavior commonly seen in social monkeys experiencing hard times:[1–9] she harasses a junior female, even killing the junior's newborn. I dislike dwelling upon the ugliness of the early diving CLCA female scene. I see a greater female dunking a lesser, obliging her to struggle for her next breath. When a male stranger approaches her congregation, I see this gathering in a tight circle to protect infants at its center, the lesser female handing in her infant for protection and failing to recover it when the danger is past. I see her kept away from better food resources. I see her, accepting defeat and like the harassed monkey female of low rank, attaching herself to a mother in need of help.

As a new species, *Sahelanthropus tchadensis*, takes shape, the junior female's attachment matures to produce a sociosexual alliance whose primary object is to see a mother's daughters to sexual maturity. Evolving in support of the alliance, a female-female pair bond unites a mother and a partner. I can appropriately call the latter a protolesbian — a behaviorally bisexual female experiencing a special attraction toward a certain mother. The protolesbian is our line's first uniquely human altruist — the first hominin deserving the adjective "selfless."

The female-female pair bond is a strong presence for many women of today, even in those who feel no sexual attraction for another woman.[10,11]

In charity, I hope the protolesbian appears quickly at Lake Mega Chad. I see a possibility this may be so. Someone whose living depends upon scores of visits daily to water-bottom is likely, on occasion, to misjudge the time it is safe to stay below. This likelihood is all the greater when she is learning to dive. Some members of a congregation are half-sisters. Many sister-pairs, I suggest, look out for each other. An older sister may "adopt" a younger and teach her to dive. An early protolesbian-mother pair, indeed, may join a younger to an older — or (what may astonish you) an older to a younger. In some social primates, it is often a younger sister who, with an older sister's help, advances to alpha-female status.[12]

Ubiquity of female-female sex in non-human primates encourages me to postulate that a protolesbian-mother pair often engages in sex for both pleasure and reinforcement of their bond. Many of you have surely seen popular accounts of the high incidence of female-female sexual activity in the bonobo (a.k.a. pigmy chimpanzee), an activity often seeming both a source of pleasure and means for cementing female cooperation in a non-kin congregation that succeeds in dominating its males.[13,14] Female foreplay has been described for a number of other social primates.[15-17] Often, in these species, a fertile female engages in sexual activity with a second female for tens of minutes, achieving external physiological signs of clitoral orgasm. At times the activity appears to be a preparation for the first female's engagement in a far briefer sexual encounter with a male. In men of today, a relic of such encounters in *Sahelanthropus* is that, typically, pornography depicting female-female sex is more arousing sexually than images of female-male sexual activity.[18,19] Their distant male ancestor, watching two females excite themselves, knows his turn is next — indeed, perhaps two turns in quick succession.

Since the protolesbian often holds her lover's infant, she is obliged to copulate with all males of the couple's guard in order to ward off danger to the infant from a male who has not known her. A protolesbian must worry lest even her guard's most junior member might kill an infant-in-arms if she has not sufficiently often engaged him in sex. Endowed with an internal signal of her fertility and no external sign of it, she can avoid male sex when fertile. At such times, however, she needs relief of an itch on the front wall of her vagina,[20] a relief that her lover's finger can provide. (In the modern woman, the G spot is a relic of this itch.[21,22]) The mother, in turn, surely develops female climax awareness. For the pair, the protolesbian's climax becomes a "sign" of their bond. When infertile, the protolesbian may enjoy a pleasure that an erect penis can provide better than a finger.

The protolesbian-mother pair's sexual behaviors provide means for birth control, restricting the hominin population. Raising a daughter at shore-side cannot be easy. If a mother were to bear several daughters in a row, she may postpone her next pregnancy, her protolesbian partner providing relief during intervals of fertility. *Homo sapiens* is a product of extreme K-selection, typically acting on a line of descent that inhabits a stable environment (or a sequence of such environments) offering, like Lake Mega Chad, foods in reliable plenty.[23] A K-selected species tends to occupy its ecological niche fully, most of the time; the species faces small risk in doing so. Our line's progress toward extreme K-selection commences at the shores of Lake Mega Chad.

Each generation of early hominin females makes a large intellectual investment in knowledge of a local water bottom. They must learn where water-foods can be harvested and at what times of the year. Females pass their knowledge, valuable "property," to their daughters. In females of our line, a genetically enabled impulse to nurture and teach preadolescent females appears.

As time passes, the entire female congregation acquires a genetically enabled impulse to help another female in trouble — an impulse conducive to formation of non-kin protolesbian-mother pairs — an impulse leading in our line's evolution to important features of its eventual ultra-sociality.

Adolescent petitioners and
senior male selective fostering

Members of a male buddy group are non-kin. For the male adolescent, driven from the scene of his childhood by the local male guard, much rests upon what non-kin male outsiders see in him. Denial of his petition to join their group consigns him to an outback, with no hope for sexual congress with a fertile female, ever.

How do gatekeepers of an outsider band view an adolescent newcomer? Their urgent need is to increase their band's fighting strength. Oldsters need younger males to fight by their side or, better, out in front, where carnage is greatest. Gatekeepers welcome a recruit promising large size in maturity

and cool courage in face of danger or pain. They subject an
arrival to a hazing that life at today's all-male military college
can mimic only at great distance.[24] Like a second-year student
at such a college, when a cadet achieves band membership, he
cheerfully becomes a primary agent for testing a more recent
arrival.

In time, some mature males **select** certain juniors for
special care: grooming for leadership or training in skills useful
in battle. Some seniors develop an impulse to teach a promising
adolescent male. A senior shows an awkward recruit better how
to throw or how to sharpen a rock. He appreciates and helps a
recruit the more who is a quick study: for the latter, educability
is an advantage. In taking on a pupil — selecting him — a
senior launches an evolutionary process, selective fostering,
that in time helps shape the human male in both morphology
and behavior. Besides educability, seniors appreciate in a cadet
those signs of worth that promise certain qualities in his
maturity: inclination to cooperate; impulse to ally with a best
buddy; loyalty to both band and buddy; readiness in battle to
risk life for either.

Australopithecines

Abundance of australopithecine female fossils is evidence that a significant change in sociosexual intercourse emerges in the australopithecine: females are now free to travel at significant distances from their waterside homes. Suggesting they do so under protection of a male or two from their guard are the fossil remains of a "first family," a group of some 13-odd males, females, and juveniles who die together in a catastrophic event (a flash flood?).[25] "Complete [australopithecine] abandonment of arboreal locomotion"[26] speaks to a change in male diet, probably greater consumption of meat. In latter australopithecine times, the guard provides flesh to their female charges: butchering sites and butchering tools appear in the archeological record at ~2.5 mya.[27,28] Yet the australopithecine buddy-group guard still expels the adolescent male. In australopithecine species, teeth counts, large and small, suggest that only about one-half of males achieve maturity.[29] Some cadets, failing to find a place in an outsider male band, do not acquire the knowledge and male companionship needful for safety from predation. Notably, teeth counts for *Lufengpithecus lufengensis* suggest equal numbers of the two sexes[30] — the situation perhaps for the harem-master/harem sociosexual arrangement when a harem-master, like the silverback gorilla, suffers presence of male juniors in his troop until, as they mature, they are about to threaten his social position.

Homo

In *Homo* species, female bodies become larger, sexual dimorphism in body size becoming less, bespeaking decline in male/male competition for access to fertile females. The adolescent male enjoys a better break than before: fossil teeth counts, large and small, are roughly equal in *Homo*.[29] Progressively larger brains bespeak both larger social groupings[31] and greater male access to aquatic foods rich in omega-3 and omega-6 fats,[32–34] which either female divers share in exchange for land foods or male imitators harvest.

I suggest a prompt for evolution of early *Homo*: the female itch diminishes; females lose touch with their interval of fertility.

From a study of teeth from 768 ancient hominin skulls, Rachel Caspari and colleagues[35] concluded that the life span of early *Homo* is longer than that of australopithecine: only 1 in 10 australopithecine achieves an age twice that for sexual maturity, while 1 in 4 early *Homo* do. (Wood and Collard[36] wrote that early *Homo* life span is roughly double the australopithecine.) In a comparison of human and chimpanzee female life histories,[37] a striking feature is that the average number of childbearing years is about the same, yet for the human female, both childhood/adolescence and survival beyond menopause are much longer than these intervals for the chimpanzee. (Some chimpanzee females live as many as 10 years beyond menopause, but they do not help their daughters by grandmothering.) In my hominin story, a diving mother's difficulties are mediated by her protolesbian partner, whose attributes are essential for the well-being of a congregation of females who must swim and dive for their living. When females become unaware of onset of fertility, when all females are mothers, continuation of their ancient way of life rests upon extensions of life span, both early and late, enabling older children[38] and grandmothers[37] to provide weaned babies foods they cannot harvest themselves. For diving mothers, with these helps, intervals between pregnancies can be short. As Meredith Small commented,[38] human females are capable of breeding "at the rate and with the success of small mammals with short lives. The typical interval between children is two and a half years, compared with five years for chimps."

Especially significant for success of the *Homo* line are midwives who counter a development unique to it: danger in birthing from increase in size of the neonate's brain.

In australopithecine and earlier times, swimming females know when they are fertile, and can select fathers of their offspring. In consequence, hominin male fitness has always been lumpy, some males siring many, other males, none. In *Homo*, villages appear in which "successful," dominant men openly enjoy high fitness. Although other men do not, yet they participate in village life and cooperate in warfare protecting their village from outsiders. Most men, in their home scene, curb their ancestral instinct to kill a nursing infant; their village banishes the few who do not to a dangerous outside. (The instinct comes into play when men raid a village of an enemy band.) Typically, each successful man bonds with several women, helping raise their children to adolescence.[39] Several villages (often three) typically exchange adolescent males on the brink of maturity.[40-42]

Human female exogamy and male social dominance are recent phenomena; men form controlling coalitions when next season's food supply becomes problematic,[43] either for a farming or herding culture from uncertainty in weather or for a gathering-hunting tribe driven by agriculturalists from good land to poor.

A development in *Homo* is capability of running at speed.[44] A puzzle in human evolution is how and when our line acquires sweat cooling, the most effective means available to a land animal for dissipating heat generated by physical activity[45] (although sweat cooling requires an expensive intake of water). We are the only predator capable of persistence hunting;[46,47] a man can run down antelope or kangaroo to exhaustion. The *tchadensis* female, needing to become streamlined, not only acquires a straightened body but also loses her ancestor's fur. When does the hominin male lose his fur (a requirement for sweat cooling)? When does he first discharge a watery, salty sweat? Is this after men first imitate their sisters' dog paddle, then invent speed swimming (needing fur loss for better stream-lining), and cooperate in large parties corralling and slaughtering dugong in the Red Sea (needing sweat to eliminate salt taken in with inadvertently swallowed sea water)? Perhaps studies of human DNA can suggest answers to these questions.

A gloss on the so-called "Aquatic Ape Theory"

My lives-in-two-worlds hypothesis is, of course, a gloss on the AAT, a theory long reviled by paleoanthropology. Following Dart's discovery of the skull of the Taung child,[48] for seven decades a savanna hypothesis of hominin origins dominates paleoanthropological thought in South Africa. After WWII, when the Piltdown skull is exposed as a fake, the hypothesis becomes entrenched in both professional and popular thought elsewhere as well. Writers on human evolution seldom address two items of evidence contrary to the hypothesis:

(a) From molecular biological studies, George Todaro[49] argued that "the ancestors of man did not develop in a geographical area where they would have been in contact with the baboon." In early hominin times, the baboon is the dominant primate species throughout East Africa. In these times, baboons are releasing an infectious type C retrovirus. Todaro found segments of this virus in the genome of each African monkey and ape species, but not in Asiatic apes nor in humans.

(b) Robert D. Martin[50] rejected the savanna hypothesis on energetic grounds. The brain is an energy-demanding organ. It's a bit less than 2% of body weight yet its maintenance requires nearly 20% of energy from diet. Food supply on a savanna is too unstable, Martin argued, to support evolution of the human brain. He asked, where do hominins find realiable sources of foodstuffs, year after year, century after century, during the million-odd years of our large brain evolution?

Somewhere in Asia, Todaro suggested, is the home of certain (early? late?) hominins in our direct line of descent. An alternate possibility is a home on an island off Africa, in the Red Sea: Eritrea's Danakil Alps is an island between ~7 and 3.5 mya.[51] Did a party of *tchadensis* reach Danakil Island at ~6 mya? Its ancient strata deserve attention from paleoscience: do they hold fossils of smaller-than-usual herbivores, a common feature of islands?[52–56] Do they preserve fossils of hominins with a taste for meat, developed from running down tiny ruminants — or from practice of a hunt for dugong in the Red Sea?

David Yesner[23] answered Martin's question. A coastal group experiences negligible change in abundance or kinds of waterside foodstuffs over a time scale in centuries. Landside droughts and food scarcities have no significant effect upon the group's welfare.

In 1998, Phillip Tobias, a distinguished South African paleoscientist, announced the failure of the savanna hypothesis paradigm.[57] In the 1990s, discoveries of hominins older than australopithecines made him appreciate that woodland, not savanna, is home to the earliest hominins. Australopithecines, dwelling nearby savannas (and also close to water), enter the hominin story after deterioration in World climate greatly expands East African savanna and desert, causing paleoscience to badly misjudge how and where these creatures spend their time.

The savanna hypothesis being ascendant in 1960, insiders of paleoscience do not entertain Sir Alister Hardy's suggestion[58] that our ancestors are "more aquatic in the past." Not an expert in paleoanthropology, Hardy does not argue from fossil data. Although from the early 1970s other outsiders produce a literature elaborating an "Aquatic Ape Theory," for insiders this becomes (in Tobias's words[57]) "a bit of a joke . . . conjuring up visions of a creature that spent all — or almost all — of its time in the water." The annals of science, I suspect, offer few examples of senior practitioners, like Tobias, who abandon a paradigm central to their own and their teachers' thought.

Evidence for the AAT goes beyond anything I have written so far. We share with marine mammals features that, in us, are unique among primates.[59-63] Most telling is the plumpness of the human newborn. Immersed suddenly in deep water, a newborn is buoyant and happy; it swims reflexively, with open mouth and closed glottis keeping water from its lungs. Like the sea otter, the infant can sleep, safely, for hours floating on its back. Even to the age of ~2, a child can hold its breath an astonishingly long time — a capability that evolves to protect it from drowning and not to blackmail its parents once the infant discovers how alarmed they become when it practices the stunt.

Evolutionary role of selective fostering

I suggest that male-male selective fostering has played — and still plays — an important role in the story of our line.

A peculiarity of the story is how many of our qualities (behavioral and morphological) have been shaped by neoteny — nee • *uh* • **tee** • nee — retention in the human adult of an attribute or quality of the human child or presence of a morphological feature of a distant ancestor's juvenile.

● For example, humankind's love of story and theater is an extension into adulthood of the child's love of fantasy, of building castles in Spain. As we will later see, extension of a childhood quality into adulthood becomes an ingredient in humankind's curious sexualities.

● The flat face of today's *sapiens* resembles that of a chimpanzee fetus early in its gestation. Centrality of the foramen magnum is a primitive feature in the mammalian Order; its evolution in the *S. tchadensis* female is an example of neoteny.

Neoteny plays almost no role in the chimpanzee's evolution. Why is it so important in ours?

Suggesting an answer is a brave Russian experiment begun in 1959 by Dimitry Belyaev — bell • off — and continued by Lyudmilla Trut. They bred foxes selecting for "tameability" — for friendliness with humans, for absence of fear of humans.[64,65] At each generation, they divided the foxes into four types: little altered; easier to handle but showing no attachment to a handler; friendly to handlers; beyond friendly, seeking human contact — indeed, much like dogs. At about the ninth generation, physical changes appeared: white fur patches, floppy ears, curved tails. As the experiment progressed, the proportion of "beyond friendly" increased.

The work discloses a strong connection between selection for behavioral traits and morphological development. Genetic transformations affecting behavior entrain genetic events. After ~40 generations, many of Trut's "tame" foxes show a white patch at center of forehead, a raised, curled tail, brown mottling, gray hairs, floppy ears, short tail, "feminized" head (shorter and wider). They resemble dogs in behavior — not only eager for human company and approval but as well possessing the dog's "social intelligence," enabling them to read human body language.

Josep Call[66] asked, referring to creation of dogs through domestication of wolves, "Is it similar to what happened to us?" A reasonable hypothesis[67] is that selection pressures operate in our line similar to those of the Belyaev-Trut experiment:

● The speciation of *S. tchadensis* creates a social creature, with its female congregation and two male buddy groups (insiders and outsiders).

● In each successive hominin speciation, agents of evolutionary change select for qualities conducive to competent participation in a more complex social life.

● Eventually, *sapiens* is able to organize its affairs in an effective, even "prosperous" Upper Pleistocene gathering-hunting band, agents having selected for humankind's ultra-sociality as well as for what one might loosely call "domesticatability" — a parallel for Trut's "tameability." Selection for this quality has been **especially important in the hominin male**.

What a great distance separates behaviors of our last male ape ancestor and those of a majority of *sapiens* males.

Are not agents for change in hominin behavior responsible as well for our line's morphological gracilization, a parallel for the neotenic outcomes of the Belyaev-Trut experiment?

Female choice no doubt is an actor promoting neoteny in our story. Also important, I argue, are agents arising from our line's expanding sociality itself. Outcomes in our social arrangements are themselves responsible, in part, for further evolutionary change. Agents of social selection, often reflecting actions or choices by senior members of either dyads or groups, promote what is conducive to group welfare. They inhibit, even expunge, what is inimical thereto.

I claim a significant role for selective fostering, a selection modality arising primarily from how a social group organizes its effort. It is ancient in hominin males.

Consider the trajectory of human evolution in the recent past. After kingdoms and cities appear in the Middle East's Fertile Crescent, conquering bands sweep down from the steppes of middle Asia, taking control of the wealth and luxuries cities offer. A charismatic leader of an egalitarian force founds an empire. His great grandson, in contrast, is an autocrat, albeit often exercising his power on relatively unimportant matters. He tends to ignore much of the business of government, delegating it to a many-layered bureaucracy. This acts like a "tribe," apart from all else. Its agenda largely focuses upon its own interests including its survival against all eventualities, even its ruler's defeat by another band thundering down from middle Asia. Surely, paralleling this history is a change in the junior type whom powerful seniors select for fostering. Initially, criteria for selection are loyalty, boldness, courage, strength, skills in warfare, willingness to endure hardship — reflections of attributes of the conqueror and his lieutenants. Over time, criteria for a junior's success become skills in jockeying for power and position within the bureaucracy, in subtlety of argument, in dissembling.

Consider the present trajectory of human evolution. This may not be inexorably leading toward change that most of you would consider "good." I can wonder if some instances of dyadic fostering tend to weed out cooperators. Are these occurring in numbers sufficient to tip an evolutionary balance in favor of the selfish? Consider the trade in illegal drugs. Aside from other social ills arising from their poorly-conceived, poorly-executed prohibition, consider how, often, a fifteen-year-old boy recruits an eight-year-old to push drugs on an inner-city schoolyard, the older boy seeing in the younger a reflection of himself when he was eight.

Selective fostering plays a role in the interaction of genes and culture, contributing to rapid expansion of a new specialist sub-population occupying a new socioeconomic niche. This is so for artists and artistes of many thousands of years ago, when they begin to enjoy reproductive success and foster youth in whom they see qualities like their own. It's so today: explosive growths of information and biotechnology industries provide modern examples of new niches affording opportunities to new types. As each new niche appears, fosterers quickly emerge: organizers, innovators, committed teachers. In some instances, to exploit a new economic niche, a new sub-population must inherit a genetically complex behavioral polymorphism. Other new sub-populations, through selective fostering, draw recruits from a largely unspecialized majority. Some sub-populations are reproductively semi-isolated. Many prize qualities whose incidences selective fostering can increase: docility; obedience; deference to authority; conformism; indoctrinability; reasonable balance between selflessness and a productive level of selfishness.

My emphasis here on the senior/junior male dyad reflects my view that it plays a critical role throughout our line's story. I do not intend a value comparison between a male/male teaching impulse and the impulse a woman feels to teach her offspring of either sex. The female impulse, of course, is far older; is vital to the mammalian Order's success; and, in *sapiens*, comes to include instruction for youth other than the woman's own progeny.

In a cross-cultural study of pre-industrial cultures, Peggy Reeves Sanday[43] found substantial equality of the two sexes in

cultures that enjoy reliable supplies of food. She saw male
dominance in cultures obliged to depend upon food supplies
that often fail or vary greatly. Extreme male dominance is a late
development, appearing alongside farms and herds and the
uncertainties these novelties bring. Sanday's work goes far
toward explaining a shift, at 10-odd thousand years ago and
after, to male social dominance, female exogamy, increased
concern over paternity confidence, and importance of patrilineal
lines of descent.

In agricultural societies, extremes in male dominance
parallel rise of cities and male literacy. A factor tending to
maintain male power is the male impulse to initiate and teach
the promising cadet. With agronomy, animal husbandry,
military arts, horsemanship, metallurgy, land surveying,
governing from a distance, tax-collecting, accounting, banking,
and (especially) writing and reading what is written, the male
adolescent is better served than his sister. Even today, the
master/apprentice relationship in both engineering[68] and
science[69] may contribute to continued male leadership in these
activities.

Proposal for a research

I can propose a research casting light upon selective fostering's significance in evolution. Examine the "personalities" of several corporations and relate these to patterns in choices made by those who have over time recruited new employees. From my experience as a consultant for a number of companies, I am confident that a researcher can identify companies of widely different corporate personalities. Some in my experience are like happy families, each employee delighted when another has an important idea or achieves a useful goal. In other companies, internal competition is fierce, employees eager to harm a within-company rival whenever opportunity arises. One company of the latter type never paid a statement of mine short of six months, using my money for its working capital. I believe the persons a company authorizes to "select" new employees often display relatively homogeneous qualities — that their selective fostering of youth of like qualities influences the company's future.

References

1.   Hrdy S.B. "Care and Exploitation of Nonhuman Primate Infants by Conspecifics other than the Mother." *Adv Study Behav.* 6:101-158. 1976.

2.   McKenna J.J. "The Evolution of Allomothering Behavior Among Colobine Monkeys: Function and Opportunism in Evolution." *Amer Anthropol.* 81:818-840. 1979.

3.   van Schaik C.P., van Noordwijk M.A. "Social Stress and the Sex Ratio of Neonates and Infants among Nonhuman Primates." *Netherlands Journal of Zoology.* 33:249-265. 1983.

4   Hamilton III W.J. "Significance of Paternal Investment by Primates to the Evolution of Adult Male-Female Associations." In Taub D.M., *Primate Paternalism: An Evolutionary and Comparative View of Male Investment.* New York: Van Nostrand Reinhold. 1984. pp. 309-335.

5.   Jolly A. *The Evolution of Primate Behavior.* New York: Macmillan. 1985. pp. 269-273.

6.   Nicolson N.A. "Infants, Mothers, and Other Females." In Smuts B.B., Cheney D.L., Seyfarth R.M., Wrangham R.W., Struhsaker T.T. *Primate Societies.* Chicago IL: Univ. of Chicago Press. 1987. pp. 330-342.

7.   Parker S.T. "A Sexual Selection Model for Hominid Evolution." *Human Evolution.* 2:235-253. 1987.

8. van Hooff J.A.R.A.M. "Macaques and Allies." In Grzimek B. *Grzimek's Encyclopedia*. New York: McGraw, vol. 2. 1990. pp. 208-285.

9. Maestripieri D. "Costs and benefits of maternal aggression in lactating female rhesus macaques." *Primates*. 35:443-453. 1990.

10. Faderman L. *Surpassing the Love of Men: Romantic Friendship and Love between Women from the Renaissance to the Present*. New York: William Morrow. 1981.

11. Parish A.R. "Female Relationships in Bonobos (*Pan paniscus*): Evidence for Bonding, Cooperation, and Female Dominance in a Male-Philopatric Species." *Human Nature*. 7:61-96. 1996.

12. Hrdy S.B., Hrdy D.B. "Hierarchical Relations Among Female Hanuman Langurs (Primates: Colobinae, *Presbytis entellus*)." *Science*. 84:913-915. 1976.

13. Wrangham R.W. "The Evolution of Sexuality in Chimpanzees and Bonobos." *Human Nature*. 4:47-79. 1993.

14. Parish A.R. "Sex and Food Control in the 'Uncommon Chimpanzee': How Bonobo Females Overcome a Phylogenetic Legacy of Male Dominance." *Ethology and Sociobiology*. 15:157-179. 1994.

15. Hrdy S.B. *The Woman That Never Evolved*. Cambridge MA: Harvard Univ. Press. 1981. pp. 170-172, 236.

16. Akers J.H., Conaway C. "Female Homosexual Behavior in *Macaca mulata*." *Arch Sex Behav*. 8:63-89. 1979.

17.   Wolfe L.D. "Japanese Macaque Female Sexual Behavior: A Comparison of Arashiyama East and West." In Small M.F. *Female Primates: Studies by Women Primatologists*. New York: Alan R. Liss. 1984. pp. 141-157.

18.   Sakheim D.K., Barlow D.H., Beck J.G., Abrahamson D.J. "A Comparison of Male Heterosexual and Male Homosexual Patterns of Sexual Arousal." *J Sex Research*. 21:183-198. 1985.

19.   Weinrich J.D. *Sexual Landscapes: Why We Are What We Are, Why We Love Whom We Love.* New York: Charles Scribner's Sons. 1987. pp. 47-49, 179-181.

20.   Morgan E. *The Descent of Woman*. London: Souvenir Press. 1972. p. 86.

21.   Ladas A.K., Whipple B, Perry J.D. *The G Spot and Other Discoveries about Human Sexuality.* New York: Holt, Rinehart, and Winston. 1982.

22.   Geddes L. "The joy of the G spot." *New Scientist.* 200(2687/2688, 20/27 December):36. 2008.

23.   Yesner D.R. "Martime Hunter-Gatherers: Ecology and Prehistory." *Current Anthropology.* 21:727-750. 1980.

24.   Faludi S. "The naked Citadel." *The New Yorker.* 70(5 Sept.):62-81. 1994.

25.   Johanson D.C., Taieb M., Coppens Y. "Pliocene Hominids from the Hadar Formation, Ethiopia (1973-1977): Stratigraphic, Chronologic, and Palaeo-environmental Contexts." *Am J Phys Anthropol.* 57:373-402. 1982.

26.   Lovejoy C.O., Suwa G., Spurlock L., *et al.* "The Pelvis and Femur of *Ardipithecus ramidus* Reveals the Postcrania of Our Last Common Ancestors with African Apes." *Science.* 326:73, 100-106. 2009.

27.   Bilsborough A. *Human Evolution.* London: Blackie. 1992. p. 137.

28.   Wood B. "The Oldest Whodunnit in the World." *Nature.* 285:292-293. 1997.

29.   Oxnard C.E. *Fossils, Teeth and Sex: New Perspectives on Human Evolution.* Seattle: Univ. of Washington Press. 1987. p. 149.

30.   Kelley J., Xu Q. "Extreme sexual dimorphism in a Miocene hominoid." *Nature.* 352:151-152. 1991.

31.   Dunbar R.I.M. "Neocortex size as a constraint on group size in primates." *J Hum Evol.* 20:469-403. 1992.

32.   Broadhurst C.L., Cunnane S.C., Crawford, M.A. "Rift Valley lake fish and shellfish provided brain-specific nutrition for early *Homo.*" *Br J Nutr.* 79:3-21. 1997.

33.   Crawford M.A. "The role of dietary fatty acids in biology: Their place in the evolution of the human brain." *Nutr Rev.* 50:3-11. 1992.

34.   Crawford M.A., Marsh D. *The Driving Force: Food in Evolution and the Future.* London: Heineman. 1989.

35.   Caspari R., Lee S.-H., Goodenough W.H. "Older Age Becomes Common Late in Human Evolution." *Proc Natl Acad Sci USA.* 101:10895-10900. 2004.

36.   Wood B., Collard M. "The Human Genus." *Science.* 284:65-71. 1999.

37.   Hawkes K., O'Connell J.F., Blurton Jones N.G., *et al.* "Grandmothering, menopause, and the evolution of human life histories." *Proc Natl Acad Sci USA.* 95:1336-1339. 1998.

38.   Small M.F. "Mother's little helpers." *New Scientist.* 176(2372, 7 Dec): 44-47. 2002.

39.   Hawkes K., O'Connell J.F., Blurton Jones N.G. "Hunting income patterns among the Hadza: Big game, common goods, foraging goals, and the evolution of the human diet." *Philos Trans R Soc London B Biol Sci.* 334:243-251. 1991.

40.   Fox R. *Kinship and Marriage.* Middlesex, England: Penguin. 1967.

41.   Spielman R.S., Neel J.V., Li F.H.F. "Inbreeding estimation from population data: Models, Procedures and Implications." *Genetics.* 85:355-381. 1977.

42.   Zihlman A.L. "Women in Evolution, Part II: Subsistence and Social Organization among Early Hominids." *Signs.* 4(Autumn): 4-20. 1978.

43.   Sanday P.R. *Female Power and Male Dominance: On the Origins of Sexual Inequality.* Cambridge, England: Cambridge Univ. Press. 1981.

44.   Rolian C., Lieberman D.E., Hamill J., Scott J.W., Werbel W. "Walking, running and the evolution of short toes in humans." *J Exp Biol.* 212:713-721. 2009.

45.   Carrier C.R. "The energetic paradox of human running and hominid evolution." *Current Anthropology*. 25:483-489. 1984.

46.   Krantz G.S. "Brain Size and Hunting Ability in Earliest Man." *Current Anthropology*. 9:450-451. 1968.

47.   Bramble D.M., Lieberman D.E. "Endurance running and the evolution of *Homo*." *Nature*. 432:345-352. 2004.

48.   Dart R.A. "*Australopithecus afarensis*: The man-ape of South Africa." *Nature*. 115:195-199. 1925.

49.   Todaro G.J. "Evidence Using Viral Gene Sequences Suggesting an Asian Origin of Man." In Konigsson L.-K. *Current Argument on Early Man*. Oxford: Pergamon. 1980. pp. 252-260.

50.   Martin R.D. *Human Brain Evolution in an Ecological Context. Fifty-second James Arthur Lecture on the Evolution of the Human Brain (April 27)*. New York: American Museum of Natural History. 1983.

51.   LaLumiere L.P. "Evolution of Human Bipedalism: A Hypothesis About Where It Happened." *Phil Trans R Soc London B Biol Sci*. 292:103-107. 1981.

52.   Sondaar P.Y. "The Island Sweepstakes: Why Did Pigmy Elephants, Dwarf Deer, and Large Mice Once Populate the Mediterranian?" *Natural History*. 95(9):50-57. 1986.

53.   Simmons, A. H. "Extinct Pigmy Hippopotamus and Early Man in Cyprus." *Nature*. 333:554-557. 1988.

54.   Lister A.M. "Rapid Dwarfing of Red Deer on Jersey in the Last Interglacial." *Nature*. 342:539-542. 1989.

55.   Vartanyan S.L., Garutt V.E., Sher A.V. "Holocene Dwarf
      Mammoths from Wrangel Island in the Siberian Arctic." *Nature*.
      362:337-340. 1993.

56.   Lister A.M. "Mammoths in Miniature." *Nature*. 362:288-289.
      1993.

57.   Tobias P.V. 1998. "Water & Human Evolution." *Out There*.
      38:35,38-44. 1998. Available on line at http//
      archive.outthere.co.za/98/dec98/disp1.dec.html.

58.   Hardy A. "Was man more aquatic in the past?" *New Scientist*.
      7:642-645. 1960.

59.   Morgan E. *The Scars of Evolution: What Our Bodies Tell Us
      about Human Origins*. London: Souvenir Press. 1990.

60.   Morgan E. *The Descent of the Child: Human Evolution from a
      New Perspective*. London: Souvenir Press. 1994.

61.   Morgan E. *The Aquatic Ape Hypothesis*. London: Souvenir Press.
      1997.

62.   Verhaegen M.J.B. "The Aquatic Ape Theory: Evidence and a
      Possible Scenario." *Med Hypotheses*. 16:17-32. 1985.

63.   Verhaegen M.J.B. "Aquatic versus Savanna: Comparative and
      Paleoenvironmental Evidence." *Nutrition and Health*. 9:165-204.
      1991.

64.   Trut L.N. "Early Canid Domestication: the Farm-Fox
      Experiment." *American Scientist*. 87:160-169. 1999.

65. Hare B., Plyushina I., Iganacio N., *et al.* "Social cognitive evolution in captive foxes is a correlated by-product of experimental domestication." *Current Biology.* 16:226-230. 2005.

66. Morell V. "Going to the Dogs." *Science.* 325:1062-1065. 2010.

67. Glegg K. "Not-so-wild idea." *American Scientist.* 87:196-197. 1999.

68. Squires A.M. *The Tender Ship: Governmental Management of Technological Change.* Boston MA: Birkhäuser Boston (Pro Scientiae Viva). 1986. pp. 11-46.

69. Pickering A. "Cyborg History and the World War II Regime." *Perspectives on Science.* 3:1-48. 1995. pp. 11-12.

 "[I]n our century ... love has come to be perceived as a refinement of the sexual impulse, but in many other centuries romantic love and sexual impulse were often considered unrelated." Lillian Faderman, 1981 (see Reference 42).

Humankind's curious sexualities

Loss of the primal itch accompanying an interval of fertility renders the **Homo** *female more love-capable*

Abstract

In late australopithecine times, the primal selective fostering impulse expands in certain senior hominin males to include educating a cadet in the joy of sex; their pupils have a role in emergence of *Homo*. In the Upper Pleistocene, peripatetic gay male artists and artistes, unwittingly, pursue an alternative male reproductive strategy and achieve a high lifetime count of progeny. Their sexuality reflects a neotenic extension of the Westermarck Effect into adulthood, endowing either lifetime aversion toward qualities of the opposite sex or lifetime interdiction of sexual arousal by these qualities.

63

Sexual naiveté of cadets • Sex educators

Sahelanthropus tchadensis males array themselves in two kinds of non-kin buddy-groups: mature insiders, forming a guard that asserts sexual monopoly to a shore-dwelling congregation of females, and outsiders, both mature and immature, biding time until their fighting strength is sufficient to challenge insiders in a battle, with control of the female congregation at stake. An insider guard brooks no rivalry from a sexually maturing son of the congregation, but consigns him to an outback, where he must petition for membership in an outsider band. Typically, like other juvenile mammalian males, including the chimpanzee, he requires instruction in sex.[1]

A female chimpanzee is apt, on her own, to get it right the first time; a male youngster needs a mature female to show him how. We lack means for measuring pleasure in another species (new instruments targeting brain activity may in time fill this void); yet our ape male cousins do not **seem** to derive pleasure from coition. A chimpanzee male's courtship typically lasts less than a minute; after coupling, he reaches his climax in less than 10 seconds, having thrust his pelvis fewer than 10 times.[2-5] For him, sex seems a bit of business: while copulating, he may engage in a second activity, such as eating a banana.[6]

Early hominin males, for several million years, are likely as clueless about the joy of sex as the male chimpanzee. From earliest hominin times, females pleasure themselves through female-female sexual activity and, prepared by this, perhaps derive some satisfaction from their frequent, obligatory couplings with males of their guards. Yet, eventually, some females succeed in teaching certain insider males the pleasures of foreplay, the pleasure of pleasure postponed. These males, when defeated by outsiders in a battle and thereby exiled from female contact, require relief from sexual frustration.

An effort of imagination is required to visualize the position of the sexually naïve male adolescent of early hominin times. Consigned by his mother to her buddy-group guard when she weans him at ~4 years, his opportunities to watch female-male sex are likely far fewer than the juvenile chimpanzee's; most sexual activity between a mature female and a member of her guard occurs in private. Today's child hears from older contemporaries vivid (if at times not entirely accurate) descriptions of the full range of human sexual activity. On achieving sexual maturity, today's child has at least a fair notion of what sex is about. In contrast, the early juvenile, upon joining an outsider band, needs to be shown even how to masturbate. He needs a teacher.

In today's prison, an older inmate often adopts a newly arrived "chicken," requiring sex in exchange for protection. An older/younger male relationship of this nature emerges in outsider hominin bands. An older male demands sex from a naïve cadet.

In time, after certain mature males have learned the joy of sex, some of these males develop a new quality: a fascination with a junior male's climax. Having known climax themselves — having learned the hedonistic pleasure it can provide — they empathize with a junior's first sexual experience. Already possessing an impulse to selectively foster juniors — already having an impulse to initiate and teach juniors in activities conducive to welfare of their band — these impulses expand to include initiating and teaching selected adolescent males in sex — to introduce cadets to the pleasures of dalliance, of prolonged arousal. In a word, these seniors become what I can appropriately call protogay males. Whatever the naïve hominin pupil brings to his first lesson in sex, there is little doubt that his protogay instructor can teach him much.

Male climax fascination is one in a suite of attributes uniting in most (but probably not quite all) contemporary gay men. Some (but surely not a majority) experience an impulsive involvement with a junior's first sexual experience. Fascination with the male climax is present, in some degree, in most modern non-gay men as well as gay. Attesting thereto is male delight in fireworks, firearms, fountains, geysers, and other eruptive phenomena as well as the obligatory presence of the cum shot in video pornography vended to both majority males and the gay minority.[7] An image of a pubescent boy's rump is sexually arousing for many non-gay men.[8] There remains in them, it seems, a residue of a primal protogay fascination with a boy's first sexual stirrings. This fascination in today's man; his susceptibility for romantic response to another man; his impulse to enter an alliance with a male buddy; his impulse to identify with a band of buddies — all of these are glorious features of our humanity, worth treasuring and honoring. Their manifestations can be seen in

- the adolescent's experimentation in male-male sex;

- the best buddy who becomes jealous when his friend marries;

- cultures in which male-male sex is socially directed behavior in each man's coming of age;[9–13]

- male-dominant cultures like that of ancient Greece, whose elite glorify male-male love and male beauty in literature and art (yet hold a mature gay man in derision). John Boswell[14] noted how, in many cultures, it is the male form, not the female, that provides the conventional standard of beauty, the object of praise from poets.

The population of the last australopithecine species in our direct line, I suggest, includes a number of protogay males who expend special effort on behalf of juniors in whom they see potential for enjoyment of sex and ability to return pleasure for pleasure. This protogay cohort plays a role in evolution of the first *Homo* species. A female couples less often with the sexually naïve, oftener with a sophisticate in sex who enjoys sexual foreplay, a behavior already ancient in the hominin female. He shares her awareness of sex's hedonistic possibilities, and a coupling that provides pleasure is more likely to leave her impregnated.[15] In *Homo*, a new sociosexual arrangement reflects incidence of a female/male pair who cooperate in rearing the female's babies to adolescence. For the pair, sex is a social glue.

Choosy mothers • Secret fathers

Brown Eyes has caught sight of Bandit, even as he hides as best he can from the rest of the Addo Gang.[16] She is the Gang's alpha female, zealously defending her exclusive right to bear babies. A strong bond unites her and the Gang's alpha male, Nkosi, who takes a large responsibility for the Gang's safety from predators, meerkat neighbors itching for war, and random trampling by unheeding elephants. Yet now, when she has turned over her weaned babies to adult mentors who will teach them meerkat ways, all she allows Nkosi is a mild hug. She is ready to cooperate with Bandit in his purpose, seeing him as brave, daring, and quick of mind in face of danger. Finally, on a lazy afternoon when the Gang is resting, Brown Eyes and Bandit escape its attention; he leaves her pregnant.

Bandit's behavior is an alternate male reproductive strategy. Alternate strategists operate with success in a number of mammalian, bird, fish, crustacean, and insect species.[17–22] They cannot succeed, of course, unless they find partners who cooperate. Ubiquity of successful alternate male reproductive strategists speaks to some degree of female promiscuity across a wide range of species.

Sarah Hrdy[23] observed that "anthropoid females were selected to ensure — one way or another — that they mate with a range of male partners." Gibbon[24] and chimpanzee[25] female behaviors illustrate Hrdy's generalization: in both species, outsider males sire a majority of offspring. Also illustrating it are behaviors of contemporary women, who sometimes welcome an approach from men other than their permanent partners. A review of anthropological data for 56 human cultures[26] shows it likely that, in 32 of these, "extra-pair copulations" (EPCs) account for some pregnancies. Paternity studies in industrialized societies[27] give further evidence that alternate male strategies are ancient in our line. Women are particularly receptive to sexual overtures from men other than their mates a few days before a moment of maximum likelihood of initiation of a pregnancy (when sperm from an EPC can compete with a mate's sperm). Although paternity studies report a wide range for the percentage of human children unrelated to their mothers' mates, a likely figure is around 10 to 15%.[28,29]

Peripatetic artists & artistes

The "explosion" of arts and crafts after ~40,000 years before the present is evidence that many Upper Pleistocene communities prosper: they do not worry about tomorrow's or next year's food. They can afford to engage artists who produce cave paintings like those seen in southern France and northern Spain. Students of this art believe it employs a syntax of drawing that reflects an academic tradition.[30,31] Upper Pleistocene cultures must grant certain young men opportunity to learn a graphic art or to practice a theater art. In maturity, if successful, these men serve their peoples by painting on bodies or cave walls, sculpting body ornament or religious memento, or performing a theater art. Mature artists take on pupils to whom they pass their techniques and traditions: they afford their pupils a fostering that is a significant ingredient behind the Upper Pleistocene cultural explosion.

These developments create opportunities for a new alternate male reproductive strategist, viz., the peripatetic professional artist or artiste:

- a dancer, musician.
- a poet, teller of stories, historian.
- an acrobat, juggler, animal tamer.

- a cosmetician, fabricator of bodily adornments, sculptor, cave painter.

No matter his line of work, he finds a warm welcome wherever he travels. He is charismatic, gentle in his approach to a woman, often more "feminine" than the average male in both appearance and behavior, although, too, he is courageous and tough — his success as a traveler in a wilderness the separates human communities is evidence of these qualities. He catches attention of community's women when he arrives. Months later, when he departs, he leaves some of them pregnant. Cuckoo-like, he foists upon mainstream males the task of helping to rear children not their own.

An impediment to his alternate reproductive strategy, however, is his sexual attraction toward women. If unable to conceal it, his behavior alerts the community's men, putting them on guard.

A removal of the impediment emerges in qualities and behaviors of an artist who is gay. Note that the gay artist does not himself "select" these qualities and behaviors, by consciously adopting them. Natural selection has done the job. Endowed with a sexual attraction toward other males, he arrives at a community to find a welcome from both men and women.

Both are attracted by his art and the entertainment he can provide. Both see in him opportunity for injecting sexual variety into their lives. Fathers understand, too, from their own life histories, that a competent instructor has appeared who can initiate and teach their adolescent sons in sex.

The life of a male berdache in a once-prosperous North American Indian gathering-hunting culture[32] can well approximate that of a professional gay male artist of the Upper Pleistocene. A berdache's sexual life is open, understood by all. No shame or loss of status is attached to a man of the majority orientation whom a berdache openly chooses as a sexual partner.

Upper Pleistocene women see in the skillful gay artist an attractive target for seduction. Seduction is flattering. For some at least, response is easy. Keep in mind that many men of today are programmed to seek sex's pleasures and to supply whatever fantasy is necessary for the endeavor in a given instant. Many men find facultative sex easy when a sex partner's gender is not the one they generally prefer.

A consummated seduction places an Upper Pleistocene gay male artist in the role of alternate reproductive strategist. His lifetime count of offspring can surpass even his most successful mainstream male competitor. Notice that his success does not reflect a conscious strategy. His (often) feminine qualities, artistic activity, lack of sexual interest in females, interest in males, inclination to initiate and teach an adolescent male youth in sex — these attributes help a woman escape notice when she engages him in a seduction. Could there be better camouflage for a human male alternate reproductive strategist? Especially against his own self-knowledge?

Note that I do not postulate high reproductive fitness for **all** gay men of the Upper Pleistocene. I have mentioned the lumpiness of male fitness in polygynous species.[33] The lumpiness extends to the sexual minority as well as to the majority. Some gay men who fail as artists or have no inclination in this direction remain bachelors and leave few descendants. Others ally with women and sire offspring. For some women, their gentle disposition and feminized features give them advantage over other men in a competition for mates.

Introduction to gay male "type"

A universal attribute of gay males is sexual interest in males; at least in the sexually naive, it seems, there is absence of such interest in females.[8,34,35] Respecting other attributes, Frederick

Whitam studied gay male communities in several cultures of very different erotic traditions: Anglo-Saxon, Afro-Brazilian, Latin, Central American Indian, and Southeast Asian.[36] In each of these diverse traditions, he saw

- a preponderance of gay men who reported cross-gender behaviors and fascinations in boyhood or whose playmates called them "sissy";

- a majority of gay men with strong interest in dance, theater, entertainment, decoration, or embellishment, and relatively low level of interest in athletics; and

- a minority of gay men who enjoyed cross-dressing.

In each culture, he identified "three broad groupings" of gay men:

- **A first grouping,** ~25% of gay men, who exhibit "extensive early cross-gender behavior as children, and as adults remain more or less effeminate in their overt behavior, continuing to manifest some cross-gender behavior."

- **A second grouping**, ~65% of gay men, who are, as children, "cross-gendered to varying degrees, but as adults are overtly masculine."

● **A third grouping**, ~10% of gay men, who manifest "no early cross-gender behavior and as adults are quite masculine in overt appearance."

Whitam's data suggest strong correlation — a genetic linkage? — between the gay male sexual orientation and a keen absorption in the arts. Whitam's work documents a significant gay male presence in professions related to the arts. A large literature supports this and other aspects of Whitam's study.

Gay men of Whitam's first grouping furnish a homophobic culture its gay stereotype. The grouping's members, typically, enter "gay" occupations in fashion, decoration, grooming, entertainment, nursing.

Members of the second grouping, in comparison with the male majority, often display far more interest in performance and visual arts, and some are professionally involved with one of these arts; but not all. Many take up occupations traditionally considered masculine.

Most studies of male homosexuality have focused upon what I may call the "standard" gay male: the first of Whitam's groupings along with part of the second. A standard attribute is a feminine style: in interests, posture, walk, voice, personal decoration, level of aggression, even shape of body. Few studies (if any?) have taken much note of the thoroughly masculine subjects of Whitam's third grouping. Any suggestion of the etiology of male sexual orientation needs to take this grouping into account: gay men exist in many kinds.

How many of today's gay men have responded to an advance by a woman? How many marry and openly sire offspring? Answers to such questions are not easily come by. Some men are not readily accessible to a survey of sexual behavior.[37]

Sex on the mind

I must now say something about love and lust — hoping to demystify the former by a little and to fasten a bit of respectability onto the latter.

An ability to share dreams, via language, is the crux of *Homo sapiens*'s astonishing success. Collectively, we profit from outcomes of dreaming: in myth, art, ethics, science, and technology. Another unique feature in us — or shared with *Orcinus orca* and other cetaceans? with elephants? — is our inner voice in combination with rich capabilities for communicating through speech. We talk to ourselves, conducting an inner debate that we like to call "thinking."

At some point in hominin evolution, our line acquires a "story" for love. Behaviors of primeval pairs of female lovers reflect a wordless "myth" that "explains" their emotional situation. A "story" does not need to be told to become a myth. Love does not need language.

The female of an early *Homo* species develops a tendency to love a certain male, and she bonds with him in a child-rearing alliance.[33] Whether she limits her sexual activity to her mate, thereby producing full siblings (who, sharing more genes, will tend to help one another more than would half-siblings[38]), or whether she enters into EPCs with other men may depend upon how closely her major histocompatibility complex (MHC) resembles that of her partner; if too similar, she is apt to engage in an EPC.[39] The MHC genes play a major role in the immune system; for a parental pair, dissimilarity is best, producing the widest range of alleles in their child.

Literature on human evolution is replete with "explanations" for the disappearance of the itch in estrous females.[40] Dare I propose yet another? The loss makes the female more love-capable. From early hominin times, the female's body develops erogenous capabilities without parallel elsewhere in the mammalian Order. Along with a neotenically evolved impulse to cuddle, her body acquires mechanisms for sexual arousal ("lust") from physical contact with a partner. A general state of arousal is a fit ally of love. The sharp urgency of the primal mammalian itch, signaling onset of an interval of fertility,[41] is not consonant with this emotion.

Males acquire their "story" much later. The physical nature of male sexual arousal precludes male love from quite matching his consort's. A female can lose her primal itch. A male cannot ignore his erect penis. He cannot ignore an urgent tingle at its tip that, on occasion, drives him to seek quick satisfaction, giving lust a bad name.

We need not wonder at the commercial success of enterprise of our time purveying stories of romance to girls and women, nor at the success of merchants vending hard-core pornography to boys and men.

An eye for sex

A characteristic of the majority of our species is a **continuous impulsive arousal by qualities of the opposite sex** (i.e., lust for healthy, vigorous exemplars of this sex who come into view). Modern lesbians and gay men display a **continuous impulsive arousal by qualities of the same sex.** In our adulthood, we experience sexual arousal, by one of the sexes, in at least three respects:

- by our ideal of the qualities of that sex which is on our mind — we can turn ourselves on simply by summoning up this ideal in our inner eye;

- by aspects of many exemplars of this sex who display features that approximate our ideal; and

- by qualities of a certain relatively few familiars of this sex, seen at firsthand, who possess a sufficiency of what attracts us.

Notice that our ideal may be culture-laden. A costume, a perfume, a tattoo, an artificial distortion of an item of anatomy: any such may be a turn-on for one culture, yet destroy desire in another.

As most of us go about our daily affairs, our minds almost never completely ignore our continuous impulsive arousal. In our time, advertisers exploit this trait, reinforcing it in those aroused by their sexual opposites; yet for most lesbians and gay men, awareness is continuous, too, with a little help from the advertising industry.

If lust is sex specific, nevertheless, love is general. Common features of the human condition are a **general love**

for all, including the stranger, and a **romantic response to an individual person of either sex.** All love is conditional. In some degree, general love is granted or withheld in light of its effect upon our own reproductive fitness. We measure our charity to the distant stranger. We withhold it from the hated enemy. Note, too, that love for a person may have little or nothing to do with impulsive sexual arousal. Love's target may be an object of lust, but not always; as Lillian Faderman[42] pointed out, it was often not so in earlier centuries.

In the typical woman's daily thoughts, romantic response, perhaps, is more important than continuous impulsive arousal. The latter may exert more influence upon the hourly musings of the typical man. A schoolgirl develops a crush for a glamorous figure of the entertainment world, yet her lifetime peak in sexual activity is still a decade away. In contrast, a schoolboy is at the height of his sexual prowess; he has physical sex much more on his mind.

Nevertheless, human love becomes something women and men share, although perhaps not in perfect symmetry. Love becomes a source of myth arising from a story-teller's dream. Myths that survive the competition for an audience's "tell it again" often reflects an ancestor's primal situation. The hero quest echoes behavior of the son who leaves the scene of his nurture to seek his fortune. Reflecting a primal female experience is the hero who appears from afar and lifts Cinderella from the ashes — or kisses Sleeping Beauty — or foils Ortrud and Telramund.

Avoiding the familiar of childhood:
The Westermarck Effect (WE)

With few exceptions, sexually reproducing species outbreed, thereby avoiding incest. I need not review the century-old debate over the question: is the avoidance in humans a natural inclination or an artifact of culture? Westermarck proposed the former; Freud responded arguing the latter. His response complicated a topic that remains simple if one is willing, at least for purpose of rational speculation, to consider the human story seamless. It strains credulity to suppose we have lost whatever mechanism operates to prevent inbreeding in our pre-*sapiens* hominin ancestors.

Two phenomena, I will now argue, account for what Robin Fox[43] called the "Westermarck effect" (WE), an inborn protection against incestuous couplings in adolescence and adulthood.

I take up first an "aversion" mechanism, underlying a **sexual aversion to an opposite-sex childhood familiar** — WE (aversion). Studies of mice provide a clue to a mechanism endowing WE(aversion) in mammals (at least in part). Mice appear to judge degree of relationship through smell. Inbred male mice tend to mate with females whose MHC genes are dissimilar to their own. Experimentally, however, Yamazaki and colleagues[44] succeeded in raising male mice that behaved otherwise. They placed newborn male mice with foster parents differing only in their MHC genes. After being weaned at 21 days, the male mice were isolated together until maturity, whereupon they uncharacteristically preferred females of the same MHC type as their own. What seems at work here is a physiological mechanism creating, during a critical period in early development, a powerful distaste for a quality of the home scene. In ordinary circumstances, a male mouse directs his aversion against his near relatives, who populate this scene, not against the outlanders of Yamazaki's artificial experiment.

A study[45] of a highly inbred community of Hutterites in South Dakota showed that a dangerous degree of similarity in MHC genes is nevertheless rare. The love that draws a couple together appears to reflect some beneficial but unrecognized mechanism for detecting and responding appropriately to MHC similarity or dissimilarity.

In our species WE(aversion) mechanisms are not likely to account fully for the WE. I have argued that we humans experience (at least) two kinds of sexual attraction. Romantic response may account for a sociosexual child-rearing alliance; impulsive arousal can lead to sexual coupling simply for the pleasure of it. Human lust is a powerful emotion. For primates, it is new in our line: nothing in the chimpanzee male behavioral repertoire could be called a result of an emotional state resembling human lust. With its arrival in our line, sexual arousal may overwhelm sexual aversion. Whichever of our attributes are associated physiologically with aversion, these likely diminish in the course of our story; for example, an ape possesses a far better sense of smell than we do.

The primal aversion, therefore, may no longer be as strong in some (all?) humans as in earlier hominins. I propose that in us a recently evolved "interdiction" mechanism joins the ancient aversion mechanism as a second attribute underlying the WE. The new mechanism, WE(interdiction), is an evolutionary novelty unique to our line, dealing with humankind's lust. I propose that substantially all humans are endowed with an **interdiction of arousal by opposite-sex familiars of childhood.** By "familiars" I mean to include parents as well as siblings; also, aunts, uncles, and cousins, if known on terms of familiarity.

For avoidance of incest, WE(interdiction) increases in importance as the human story unfolds. New circumstances increase risk from inbreeding. Extended parental care of adolescents beyond puberty; extended association of mature offspring with parents; emergence of grandparenting — all of these increase the odds that two closely related individuals will experience a mutual sexual attraction. In addition, commercial dealings between nearby villages require their members to visit back and forth, reuniting closely related pairs again and again. Incest, of course, does occur. Its incidence appears to reflect a lesser degree of familiarity between older and younger sex partner in the latter's childhood: father/daughter incest occurs much oftener than mother/son; sibling incest is rare.

Extension of WE into adulthood

By the time most of us reach sexual maturity, the WE no longer acts to inhibit a sexual relationship with a new familiar of the opposite sex. A regulatory "gene prompt" has turned off the WE's genetic underpinnings.

I hypothesize that in a few of us, responding to a neotenic evolutionary change, this gene prompt fails to act. In these few, the WE continues to function beyond adolescence. In other words, I postulate an extended Westermarck effect (EWE). Paralleling WE(aversion) and WE(interdiction), I postulate an EWE(aversion) and an EWE(interdiction). The former endows **lifetime aversion toward qualities of the opposite sex**; the latter, **lifetime interdiction of sexual arousal by these qualities.**

I postulate that either EWE(aversion) or EWE (interdiction) is responsible (or both are) for qualities and behaviors of significant numbers of lesbians and gay men. I emphasize that either of the EWEs is a product of neotenic evolutionary change; recall that such change in our line seems a result of selective fostering and female choice, each favoring evolution of gentler, "domesticatable" men.[33] A deep underlying "cause" for a minority sexual orientation in men could be an overreach of selection mechanisms that eventually, after several millions of years, prepare men for entering fitness-enhancing bonds with women. Such overreach not only underlies an EWE but could also account for linkage, in many gay men, between the EWE and a neotenically endowed fascination with the arts (a survival in an adult of the child's real or imagined building of castles in Spain). In many, an EWE could also be linked with neotenic genetics endowing the "feminized" qualities of the "standard" gay male. Absence of such linkages could account for Whitam's third grouping of gay men.

     I note that "incest" does not embrace same-sex relationships. Same-sex sibling or cousin pairs, aunt-niece/ uncle-nephew pairs, even mother-daughter/father-son pairs — a mechanism hindering their formation would serve no biological purpose. Someone familiar with the lesbian or gay scene will know at least a few such relationships.

Introduction to a search for "type"

An important test for my hypotheses could be an attempt to discover individuals endowed with only EWE(aversion) or EWE(interdiction) and to distinguish between their behaviors. Some lesbians seem to develop strong antipathy (even hatred) toward men; some gay men seem to shun (or hate) women. In these lesbians and gay men, perhaps only EWE(aversion) is expressed. I believe these to be minorities in lesbian and gay populations, but I am eager to see data testing my belief. In contrast, many lesbians enjoy men companions of either majority or minority sexual orientation. Many gay men develop strong associations with their women friends and even sufficient love for particular women to sustain lifetimes of child-rearing partnerships.

References

1.   Ford C.S., Beach F.A. *Patterns of Sexual Behavior*. New York: Harper. 1951. pp. 194-195.

2.   Davenport R.K. "Some Behavioral Disturbances of Great Apes in Captivity." In Hamburg D.A., McCown E.R. *The Great Apes*. Menlo Park CA: Benjamin, Cummings. 1979. pp. 341-397.

3.   Tutin C.E.G., McGinnis P.R. "Chimpanzee Reproduction in the Wild." In Graham C.D. *Reproductive Biology of the Great Apes: Comparative and Biomedical Perspectives*. New York: Academic Press. 1981. pp. 239-264.

4.   Ghiglieri M.P. *Chimpanzees of the Kibale Forest: A Field Study of Ecology and Social Structures*. New York: Columbia Univ. Press. 1984. p. 159.

5.   Jolly A. *The Evolution of Primate Behaviors*. New York: Macmillan. 1985. p. 287.

6.   Tanner N.M. *On Becoming Human*. Cambridge, England: Cambridge Univ. Press, 1981. p. 155.

7.   Faludi S. "The money shot." *The New Yorker.* 71(30 Oct.):64-87.
        1995.

8.   Freund K. "Male homosexuality: An analysis of the patterns." In
        Loraine J.A. *Understanding Homosexuality: Its Biological and
        Psychological Bases.* St. Leonardgate, Lancaster, England:
        Medical and Technical Publishing Co. Ltd. 1974. pp. 25-81.

9.   Flacelière R. *Love in ancient Greece.* New York: Crown. 1962.

10.   Dover K.J. *Greek Homosexuality: Updated with a new Postscript.*
        Cambridge MA: Harvard Univ. Press. 1989.

11.   Davidson J.N. *The Greeks and Greek Love: A Bold New
        Exploration of the Ancient World.* New York: Random House.
        2007.

12.   Herdt G.H. *Ritualized Homosexuality in Melanesia.* Berkeley CA:
        Univ. of California Press. 1984.

13.   Herdt G.H. *Guardians of the Flute: Idioms of Masculinity.*
        Chicago IL: Univ. of Chicago Press. 1994.

14.   Boswell J. "Homosexuality in historical perspective." Lecture,
        Univ. of North Carolina at Greensboro. 1982(Dec. 7).

15.   Shackelford T.K., Weekes-Shackelford V.A., LeBlanc G.J., *et al.*
        "Female Coital Orgasm and Male Attractiveness." *Human Nature.*
        11:299-306. 2000.

16.   Attenborough D. "David Attenborough's Natural World: Meerkats
        Divided." Blue Ridge Public Television, Roanoke VA. 1996
        (June 6).

17.   Crawford C.B., Anderson J.L. "Sociobiology: An environmentalist discipline?" *American Psychologist.* 44:1449-1459. 1989.

18.   Moore M.C. "Application of Organization–Activation Theory to Alternative Male Reproductive Strategies: a Review." *Hormones and Behavior.* 25:154-179. 1991.

19.   Conniff R. "Close Encounters of the Sneaky Kind." *Smithsonian.* 14(4, July):66-70. 2003.

20.   Bass A.H. "Shaping brain sexuality." *American Scientist.* 84:352-363. 1996.

21.   Gross M.R. "Alternative reproductive strategies and tactics: Diversity within sexes." *Trends in Ecology and Evolution.* 11:92-98. 1996.

22.   Neff B.D. "Something Fishy in the Nest." *Natural History.* 113(1, February):46-50. 2004.

23.   Hrdy S.B. "Raising Darwin's Consciousness: Female Sexuality and the Prehominid Origin of Patriarchy." *Human Nature.* 8:1-49. 1997.

24.   Reichard U., Sommer V. "Group Encounters in Wild Gibbons (*Hylobates lar*): Agonism, Affiliation, and the Concept of Infanticide." *Behavior.* 134:1135-1174. 1997.

25.   Gagneux P., Woodruff D.S., Boesch C. "Furtive mating in female chimpanzees." *Nature.* 387:358-359. 1997.

26.   Smith R.L. *Sperm Competition and the Evolution of Animal Mating Systems.* New York: Academic Press. 1984.

27. Bellis M.A., Baker R.R. "Do Females Promote Sperm Competition? Data for Humans." *Animal Behaviour.* 40:997-999. 1990.

28. Baker R.R., Bellis M.A. *Human Sperm Competition: Ejaculate Adjustment by Males and the Function of Masturbation.* London, England: Chapman and Hall. 1995.

29. McKnight J. *Straight Science? Homosexuality, Evolution and Adaptation.* London, England: Rutledge. 1997.

30. White R. *Dark Caves, Bright Visions: Life in Ice Age Europe.* New York: W.W. Norton. 1986. pp. 156-159.

31. Curtis G. *The Cave Painters: Probing the Mysteries of the World's First Artists.* New York: Knopf. 2006.

32. Williams W.L. *The Spirit and the Flesh: Sexual Diversity in American Indian Culture.* Boston: Beach Press. 1986.

33. Hawkes K., O'Connell J.F., Blurton Jones N.G. "Hunting income patterns among the Hadza: Big game, common goods, foraging goals, and the evolution of the human diet." *Philos Trans R Soc London B Biol Sci.* 334:243-251. 1991.

34. Freund K., Scher H., Chan S., Ben-Aron M. "Experimental analysis of pedophilia. *Behavioural Research and Therapy.* 20:105-112. 1982.

35. McConaghy N. "Heterosexual Experience, Marital Status, and Orientation of Homosexual Males." *Arch Sex Behav.* 7:575-581. 1978.

36.   Whitam F.L., Mathy R.M. *Male Homosexuality in Four Societies: Brazil, Guatemala, the Philippines, and the United States.* New York: Praeger. 1986.

37.   Bagley C., Tremblay P. "On the Prevalence of Homosexuality and Bisexuality, in a Random Community Survey of 750 Men Aged 18 to 27." *J. Homosexuality.* 36(2):1-18. 1998.

38.   Peck J.R., Feldman, M.W.   "Kin selection and the evolution of monogamy." *Science.* 240:1672-1674. 1988.

39.   Garver-Apgar C.F., Gagestad S.W., Thornhill R., *et al.* "Major Histocompatibility Complex Alleles, Sexual Responsivity, and Unfaithfulness in Romantic Couples." *Psychological Science.* 17:830-835. 2006.

40.   Hrdy S.B., Whitten P.L. "Patterning of Sexual Activity." In Smuts B.B., Cheney D.L., Seyfarth R.M., Wrangham R.W., Struhsaker T.T. *Primate Societies.* Chicago: Univ. of Chicago Press. 1987. pp. 370-384.

41.   Morgan E. *The Descent of Woman.* London: Souvenir Press. 1972. p. 86.

42.   Faderman L. *Surpassing the Love of Men: Romantic Friendship and Love Between Women from the Renaissance to the Present.* New York: William Morrow. 1981.

43.   Fox R. "Sibling incest." *British Journal of Sociology.* 13:128-150. 1962.

44. Yamazaki K., Beauchamp G.K., Kupniewski D., *et al.* "Familial Imprinting Determines H-2 Selective Mating Preferences." *Science.* 240:1331-1332. 1988.

45. Small M.F. "Love with the Proper Stranger." *Natural History.* 107 (7):14,16-19. 1998.

Acknowledgement

Having kindly read drafts of related matter, David Andrew, Richard Burian, Joseph Carroll, Kathryn Coe, the late Dudley Duncan, Robin Fox, John Hartung, Sarah Blaffer Hrdy, David Hull, Elaine Morgan, the late John Pierce, Daniel Rancour-Laferriere, Robin Russell, Nic Tideman, and Bruce Wallace gave much-appreciated criticism, comment, suggestion, or encouragement.

Biography

Although my undergraduate degree was in chemistry (Missouri University, Columbia) and my PhD in physical chemistry (Cornell University, mentored by the late John Kirkwood), I became a chemical engineer during four years on the Manhattan Project under the mentorship of the late Manson Benedict. First working for The M. W. Kellogg Co., and later for The Kellex

Co. (a Kellogg subsidiary), I participated in design and construction of K-25, the gaseous diffusion plant at Oak Ridge TN enriching U-235 from 0.7% in natural uranium to higher concentrations. From February 1945, I advised Union Carbide Co. (K-25's operator) and the U. S. Army as to what levels of concentration of U-235 it would be sensible for a partial K-25 plant to ship, and in what amounts (stages of the gaseous diffusion cascade were installed on a schedule lasting into late August). In September I helped Carbide organize a Process Analysis Department. Cuthbert Daniel, a Department member, and I instituted a statistically controlled material balance. This became important in supporting a decision to increase the concentration level in K-25's product from 35% to 90% (bomb grade) without danger of accumulation of a critical mass of U-235 creating a nuclear chain reaction within the K-25 high-end stages. The increase permitted the Army to stop using expensive electromagnetic separation to carry K-25's product to bomb level. In 1950, during visits to other sites operated by the Atomic Energy Commission, I learned to my surprise that no other site was conducting a material balance of its dangerous fissionable material therein under even remotely comparable statistical control.

Believing nuclear electricity would be too
dangerous for the often careless, forgetful human animal
to use — that nuclear technology's only viable
applications would be military, I left the field in
mid-1946. A major focus of a further 21 years of service
to industry (the final 8 years as a self-employed
consultant) was in technologies related to fossil fuels. I
then served 19 years on chemical engineering faculties
(nine at City College of New York and ten at Virginia
Polytechnic Institute & State University). My research in
those years related to control of emissions from
utilization for coal. I am still active on this topic, having
filed for three patents in the past year. I am a member of
the National Academy of Engineering, a fellow of the
American Academy of Arts & Sciences, and a fellow of
the American Association for the Advancement of
Science.

Always an avid reader of anthropological
literature, I became curious about human evolution and
have now devoted a number of years to its serious study.
A member of the Human Behavior & Evolution Society
and the International Society for Human Ethology, I have

presented papers at meetings of both societies. I have a
manuscript in progress, *The left hand of love*, which will
offer an account of human evolution in greater detail than
to be found herein.

I also developed an interest in management
through puzzlement at the several major, expensive
failures of United States governmental efforts to advance
technology following WWII. Both *Nature* and *New
Scientist* favorably reviewed my 1986 book, *The Tender
Ship*, in which I tried to develop criteria for recognizing
good or bad governmental management of technological
change.